# Genome Exploitation

## Data Mining the Genome

# Genome Exploitation

## Data Mining the Genome

Edited by

**J. Perry Gustafson**
*USDA-ARS*
*University of Missouri*
*Columbia, Missouri*

**Randy Shoemaker**
*USDA-ARS*
*Iowa State University*
*Ames, Iowa*

**John W. Snape**
*John Innes Centre*
*Norwich, England*

 Springer

Library of Congress Cataloging-in-Publication Data

---

Gustafson, J. P., R. Shoemaker, J. W. Snape
    Genome exploitation : data mining the genome / J. Perry Gustafson, R. Shoemaker, J. W. Snape.
    p.  cm.
    Includes bibliographical references and index.
    ISBN 0-387-24123-X
    1. Genomics.   2. Data mining.   I. Title.

QH452.7.G87 2005
572.8'6—dc22

                             2004065386

---

ISBN 0-387-24123-X          e-ISBN 0-387-24187-6
ISBN-13 978-0387-24123-4    Printed on acid-free paper.

Printed in Singapore    (TB/KYO)

9  8  7  6  5  4  3  2  1     SPIN 153560

springeronline.com

The editors would like to dedicate this volume to Dr. George P. Redei in recognition of his effort, support, and commitment to continuation of the Stadler Genetics Symposium. George has been the strongest supporter or the Stadler Genetics Symposium since he co-organized the very first one in May 1969 with Dr. Gordon Kimber. He continued organizing and editing the Symposium series through volume 13 in 1981. Since then he has remained tireless in his attendance and willingness to help in identifying excellent speakers.

# Acknowledgment

The editors would like to gratefully acknowledge the generous support of the following contributors from the University of Missouri: the College of Agriculture, Food and Natural Resources, the Molecular Biology Program, the Office of Research, the Interdisciplinary Plant Group, the College of Arts and Sciences, the School of Veterinary Medicine, the Plant Sciences Unit, and the School of Medicine. The contributors continued support made the 23rd Stadler Genetics Symposium a success.

The speakers, who spent a tremendous amount of time preparing their lectures and manuscripts are gratefully acknowledged. Without their expertise and dedication, the Symposium could not have taken place.

We wish to thank all of the local Chairpersons for seeing that all the speakers were well taken care of during the Symposium.

The behind-the-scene and on-site preparations were excellently handles by Jamie Schieber from Conferences and Specialized Courses, University of Missouri, who tirelessly handled all of our peculiar requirements and made sure everything was well organized.

A special thanks goes to Heather Lewandowski and Joanne Beavers, USDA-ARS, Columbia, Missouri, and Christy Trautmann, Iowa State University for all their excellent help in keeping us organized and reminding us of all the things we had forgotten.

J. Perry Gustafson
USDA-ARS
University of Missouri
Columbia, Missouri
USA

Randy Shoemaker
USADA-ARS
Iowa State University
Ames, Iowa
USA

John W. Snape
John Innes Centre
Colney Lane
Norwich
United Kingdom

# Contents

## Chapter 5

**Novel Tools for Plant Genome Annotation and Applications to**
***Arabidopsis* and Rice** ............................................................ 63

Volker Brendel

## Chapter 6

**FCModeler: Dynamic Graph Display and Fuzzy Modeling
of Metabolic Maps** ............................................................... 77

Julie A. Dickerson

## Chapter 7

**Old Methods for New Ideas: Genetic Dissection of the Determinants
of Gene Expression Levels** ...................................................... 89

Kyunga Kim, Marilyn A.L. West, Richard W. Michelmore,
Dina A. St. Clair, and R.W. Doerge

## Chapter 8

**Charting Contig–Component Relationships within the Triticeae:
Exploiting the Genome** .........................................  109

Gerard R. Lazo, Nancy Lui, Frank M. You, David D. Hummel,
Shiaoman Chao, and Olin D. Anderson

## Chapter 9

**Protein Family Classification with Discriminant Function Analysis** ..  121

Etsuko N. Moriyama and Junhyong Kim

## Chapter 10

**Exploiting Natural Variation to Understand Gene Function in Pine** .    133

David B. Neale and Garth R. Brown

## Chapter 11

**Merging Analyses of Predisposition and Physiology Towards Polygene
Discovery** ......................................................    145

Daniel Pomp, Mark F. Allan, and Stephanie R. Wesolowsk

Chapter 12

**Mining the EST Databases to Determine Evolutionary Events
in the Legumes and Grasses** . . . . . . . . . . . . . . . . . . . . . . . . . . . . . . . . . . .   163

Jessica A. Schlueter, Phillip Dixon, Cheryl Granger,
and Randy C. Shoemaker

Chapter 13

**A Biologist's View of Systems Integration Systems Biology:
The Pathogen Portal Project** . . . . . . . . . . . . . . . . . . . . . . . . . . . . . . . . . . . .   183

R. Lathigra, Y. He, R.R. Vines, E.K. Nordberg, and B.W.S. Sobral

Chapter 14

**Alignment of Wheat and Rice Structural Genomics Resources** . . . . . .   197

Daryl J. Somers, Sylvie Cloutier, and Travis Banks

*Chapter 15*

**Computational Identification of Legume-Specific Genes** ............. 211

Michelle A. Graham, Kevin A. T. Silverstein, Steven B. Cannon,
and Kathryn A. VandenBosch

## ABSTRACTS

**Brahms and Beergenes: Information Management for Genetic
Research on Barley and Oat**

Jean Gerster, Nicholas A. Tinker, Yella Jovich-Zahirovich, Anissa Lybaert,
Shaolin Liu, Stephen J. Molnar, and Diane E. Mather

*Chapter 1*

# Rice Genome Sequencing and Data Mining Resources

Baltazar A. Antonio, Yoshiaki Nagamura, Nobukasu Namiki, Takashi Matsumoto, and Takuji Sasaki

## 1. INTRODUCTION

Rice is one the most widely studied crop in terms of genetics, molecular biology and breeding science. Among the two major subspecies of *Oryza sativa*, namely, subspecies *japonica* and *indica*, a large number of local cultivars exist and allelic differences among these cultivars may underlie the adaptations that have evolved in response to a particular set of environmental conditions. The rice genome sequencing project, which has become an international collaboration, should provide the foundation for a thorough understanding of the rice plant (Sasaki and Burr, 2000). As rice is considered one of the world's most important crop having an accurate genome sequence of a single rice cultivar for use as a prototype is a requisite for identifying all genes and for associating sequence variation with phenotypic expression. Furthermore, rice is also considered a model cereal genome, which shares a common synteny with much larger grass genomes including maize, wheat, barley and sorghum (Moore et al., 1995). Thus, the complete sequence of rice will also undoubtedly be beneficial in the genomic analysis of cereal genomes in general.

**Baltazar A. Antonio, Yoshiaki Nagamura, Takashi Matsumoto, and Takuji Sasaki**   National Institute of Agrobiological Sciences, 2-1-2 Kannondai, Tsukuba, Ibaraki 305-8602, Japan. **Nobukasu Namiki**   Institute of the Society for Techno-innovation of Agriculture, Forestry and Fisheries, 446-1 Kamiyokoba, Ippaizuka, Tsukuba, Ibaraki 305-0854, Japan.

*Genome Exploitation: Data Mining the Genome*, edited by J. Perry Gustafson, Randy Shoemaker, and John W. Snape.
Springer Science + Business Media, New York, 2005.

In order to facilitate the discovery of new biological insights from the genome sequence data as well as to create a global perspective from which applied principles in crop improvement can be discerned, we are developing a robust informatics system of rice based on the genome sequence and other genomics data. These resources will be very useful for processing, elucidating and propagating genomic information in rice and facilitate data mining to extract more useful information. From the biological point of view, genomic resources derived from comprehensive sequencing will also be useful in understanding complex biological pathways as well as in uncovering the phylogenetic relationships and evolutionary patterns of cereal genomes. However, this requires new methods for analysis, tools for data mining and the development of integrated databases that would facilitate efficient access and management of different types of information. Thus a major objective of the rice genome project is to integrate the rice genome sequence data with other genomics information available for rice to allow a more efficient utilization of the entire sequence particularly in functional genomics, comparative genomics and applied genomics.

An overview of the rice genome sequencing effort and various genomic resources that have been developed as part of the Japanese program for rice genome research is described here. The URLs and major features of databases and websites generated from various projects are shown Table 1. These data mining resources will be indispensable in revealing all the unique features of rice as well as other cereal crops, which share syntenic relationships with rice.

## 2. RICE GENOME SEQUENCING

As a major food source for more than half of the world's population, the importance of analyzing the entire genome has become a research priority in the scientific community. With this situation on hand, Japan initiated the Rice Genome Research Program (RGP) in 1991 with the aim of elucidating the structure and function of the rice genome (Sasaki, 1999). During the first phase of the program, RGP has successfully established a high-density linkage map of rice (Kurata et al., 1994a, Harushima et al., 1998), an extensive catalog of rice ESTs (Yamamoto and Sasaki, 1997) and a YAC (yeast artificial chromosome)-based physical map covering the entire genome (Saji et al., 2001). These resources have been very useful in genome analysis of rice and other cereal crops. The molecular markers have been used in comparative mapping of rice varieties (Antonio et al., 1996), mapping of quantitative trait loci (Yano and Sasaki, 1997) and comparative mapping with other cereal crops (Kurata et al., 1994b). The ESTs developed by RGP from about 15 types of cDNA libraries derived from various tissues and organs such as leaves, roots, panicles and calli using Nipponbare as resource constitute the main bulk of rice ESTs deposited in public databases. The physical maps have also been useful in map-based cloning of many agronomically important traits in rice (Yano and Sasaki, 1997).

## Table 1
## Databases and Websites for Rice Genome Sequencing, Functional Genomics and Related Projects.[1]

| Website/Database | URL | Features |
|---|---|---|
| **RGP** (Rice Genome Research Program) | http://rgp.dna.affrc.go.jp/ | genomics resources such as genetic map, physical map, transcript map, genome sequence, databases etc. |
| **IRGSP** (International Rice Genome Sequencing Project) | http://rgp.dna.affrc.go.jp/IRGSP/ | sequenced PAC / BAC clones from the international sequencing consortium |
| **INE** (Integrated Rice Genome Explorer) | http://rgp.dna.affrc.go.jp/giot/INE.html | integration of genetic map, physical map, transcript map genome sequence data and annotation |
| **RiceGAAS** (Rice Genome Automated Annotation System) | http://ricegaas.dna.affrc.go.jp/ | automated annotation tool and database of annotated rice genome sequence data |
| **RAD** (Rice Genome Annotation Database) | http://golgi.gs.dna.affrc.go.jp/SY-1102/rad2/index.html | contig-oriented annotation of rice genome sequence, analysis of sequence features |
| **Tos17** (*Tos*17 Mutant Panel Database) | http://tos.nias.affrc.go.jp/~miyao/pub/tos17/ | flanking sequences with Tos17 insertions in about 5000 lines from 50000 transposon insertion lines |
| **KOME** (Knowledge-based Oryza Molecular biological Encyclopedia) | http://cdna01.dna.affrc.go.jp/cDNA/ | rice full-length cDNA sequences, annotation, homology search, mapping information, pattern of alternative splicing, protein domain, transmembrane structure, cellular localization and gene function |
| **RED** (Rice Expression Database) | http://red.dna.affrc.go.jp/RED/ | expression profiles of rice ESTs analyzed using 1265 and 8897 cDNA microarray systems |
| **RMOS** (Rice Microarray Opening Site) | http://cdna01.dna.affrc.go.jp/RMOS/main_en.html | outline of the rice microarray project |
| Rice PIPELINE | http://cdna01.dna.affrc.go.jp/PIPE/ | unification tool for rice functional genomics resources |
| RiceBLAST | http://riceblast.dna.affrc.go.jp/ | BLASt homology search using various rice genome sequence datasets |
| PLACE (Plant Cis-acting Regulatory DNA Elements Database) | http://www.dna.affrc.go.jp/htdocs/PLACE/ | motifs in plant cis-acting regulatory DNA elements |
| RMG (Rice Mitochondrial Genome Database) | http://rmg.rice.dna.affrc.go.jp/ | rice mitochondrial genome sequence and detailed analysis |
| Rice Genome Simulator | http://www.nias.affrc.go.jp/project/inegenome_e/simulator/simulator_outlook_e.htm | outline of the Rice genome Simulator Project |
| RGRC (Rice Genome Resource Center) | http://www.rgrc.dna.affrc.go.jp/index.html.en | distribution center for rice full-length cDNA, insertion mutant lines and materials for genetic analysis |
| DNA Bank | http://www.dna.affrc.go.jp/ | genome projects of the Japanese Ministry of Agriculture, Forestry and Fisheries |

[1]These websites and databases are maintained at the National Institute of Agrobiological Sciences (NIAS, http://www.nias.affrc.go.jp/index_e.html).

In 1998, the RGP embarked on the second phase of genome analysis, which was aimed at sequencing the entire rice genome. A sequence strategy based on clone-by-clone shotgun sequencing method was adopted. The genetic map, YAC-based physical map and a transcript map derived by anchoring ESTS in the YAC-based physical map (Wu et al., 2001) served as fundamental tools in constructing a sequence-ready physical map of rice. The DNA markers in the genetic map and ESTs in the transcript map were used to align large insert clones of PAC (P1-derived artificial chromosome) or BAC (bacterial artificial chromosome) clones along the 12 rice chromosomes (Baba et al., 2000). These clones were used to obtain a high-quality sequence with 99.99% accuracy.

Even with a relatively small genome size estimated at 430 Mb, the task of sequencing the entire genome is quite enormous. The genome sequencing initiative in Japan served as a stimulus for the U.S.A. as well as other Asian and European countries to establish a similar program. Eventually an international collaboration now known as the International Rice Genome Sequencing Project (IRGSP) was organized in order to share resources and accelerate the completion of sequencing (Sasaki and Burr, 2000). Several standards were decided concerning sequencing strategy, sequence quality, annotation and sequence release. The rice cultivar Nipponbare, which has been used by RGP was chosen as a common template for sequencing. This collaboration, which started with five-member countries, has expanded into a consortium of ten participating countries or regions. Each group is in charge of sequencing entire chromosomes or specific regions of a particular chromosome. The sequencing effort is also highly facilitated by contributions from the private sector. In 2000, the Monsanto Company, which conducted an independent genome sequencing of rice has made the draft sequence of more than 3000 BAC clones available (Barry, 2001). Then Syngenta, which released the draft sequence of the rice genome obtained by whole genome shotgun sequencing (Goff et al., 2002) also shared the sequence data to IRGSP.

The rice genome sequencing collaboration has made a significant harvest last year with the completion of the high-quality draft sequence of the entire genome. As of June 2003, a total of 3,349 PAC/BAC clones have been sequenced resulting in 219 Mb of phase 2 quality (all segments of sequence are in the proper orientation and order but with few gaps) sequence and 232 Mb of phase 3 quality (high quality finished sequence with no gaps) sequence. At present, chromosome 1 (Sasaki et al., 2002), chromosome 4 (Feng et al., 2002) and chromosome 10 (Rice Chromosome 10 Sequencing Consortium, 2003) have been completely sequenced by Japan, China, and the USA, respectively, at the phase 3 level. Currently the IRGSP is aiming at closing several gaps in the physical map for each chromosome as well as in raising the sequence to phase 3 finished-quality level. Ultimately the sequence of the entire genome is expected to be completed before the end of year 2004.

## 3. RICE GENOME INFORMATICS

In order to extract as much biological information from the genome sequence, the RGP has developed a robust informatics infrastructure aimed at analyzing the genome sequence, developing integrated databases and releasing the analyzed sequence through the internet. The genome sequences are initially analyzed using an automated annotation system, edited using an annotation plotting tool to construct the most plausible gene model and then incorporated into a database designed to integrate map and genome sequence information.

An automated annotation system and database called Rice Genome Automated Annotation System (RiceGAAS) has been developed to extract biologically useful and timely information from the sequence data on a regular basis (Sakata et al., 2002). The system integrates a total of 15 analysis programs for prediction protein-coding gene structure and analysis of other features of the sequence. These include GENSCAN (Arabidopsis), GENSCAN (Maize), RiceHMMand FGENESHfor gene domain prediction; BLASTXand BLASTNfor homology search against protein and rice EST databases, respectively; MZEF for exon prediction; SplicePredictor for splice site prediction; Printrepeats and Repeat-Masker for repetitive sequence detection; tRNAscan for transfer RNA prediction; HMMER ProfileScan and MOTIF for homology search against amino acid sequence motif database; PSORTfor protein localization site prediction; SOSUI for classification and prediction of the secondary structure of membrane protein; and PLACE-SignalScan for cis-element detection. The reference databases are uploaded from the original sites, stored in the system and updated on a regular basis. Thus all analyses are executed using the latest entries in respective databases. The system automatically integrates multiple analysis results and interprets coding regions using an algorithm based on the concept of combining the analysis of different gene prediction programs, homology search and analysis of other features of the sequence. Then the results are summarized in an annotation map using a web-based graphical view. The autopredicted gene set as well as the output of the different gene prediction programs and homology searches are represented as colored objects, which are clickable and provide links to the corresponding page of analysis results. RiceGAAS functions not only for analysis of RGP generated sequences but all other genome sequences generated by other IRGSP members. All rice genome sequences in GenBank are automatically collected using a keyword search executed on a daily basis against the GenBank database. Thus, all submitted sequences by participating IRGSP members are immediately annotated and the results can be viewed through RiceGAAS.

For RGP sequenced clones, the autopredicted gene sets are downloaded from the database and edited using an annotation-plotting tool to generate an accurate structure of the gene based on existing evidences. The different prediction programs use different measures to determine the coding regions so that the output may differ even for the same region of the sequence. The Combiner program originally developed by TIGR (The Institute for Genomic Research) is used to determine the best prediction output. This program integrates the results of various gene prediction programs, generates a consensus prediction, and selects a prediction output that differs least from the consensus. The gene models selected are checked against homology with protein and EST sequences. The translation products of the coding regions are examined and the exons are edited if necessary. Manual curation also facilitates identification of miscellaneous features of the sequence such as transposons and retrotransposable elements. The resulting annotation map provides the most plausible structure of all predicted genes in the sequence. Functional characterization of the predicted genes is performed by BLASTP analysis against non-redundant protein database. The predicted genes are then classified as "same protein", "putative protein", "similar to the protein", "unknown protein" and "hypothetical protein" based on the degree of similarity to known proteins in public databases.

The informatics effort of RGP centers on the database aptly named INE (INtegrated Rice Genome Explorer), which literally means "rice plant" in the Japanese language (Sakata et al., 2000). This database was developed primarily to integrate the genetic and physical mapping data with the sequence of the rice genome. In addition, it also functions as a repository of rice genome sequences from the international sequencing collaboration. At present the database consists of the genetic map with 3,267 DNA markers, a YAC-based physical map covering 80% of the genome, PAC/BAC contigs and a transcript map of rice, which consists of about 6500 ESTs. These data are integrated with the genome sequence data. The integrated maps for each chromosome facilitate a general overview of the genomic information in a particular chromosome. A relational database scheme has been implemented to improve facile access to the database and facilitate a robust searching capability. The maps can be manipulated by zoom in/out, which enables browsing at detail-oriented levels. The viewer was programmed in Java language using an application, which facilitates rapid display of integrated maps. This attribute contributes to smooth navigation of specific information associated with each data set. Furthermore, it also provides an overall view of the distribution of the markers and clones in the entire chromosome as well as specific details such as screening data, image sets, sequences, as well as detailed contig maps in the case of sequenced regions of the chromosome. Links are also provided on the curated annotation of all sequenced clones. The annotation map provides information on the structural features of the predicted genes as well the results of homology searches with protein and EST databases.

An annotation database was also developed in an effort to efficiently manage and integrate accumulated information from annotation of the genome sequence. The Rice Genome Annotation Database (RAD) system allows merging the annotation of individual PAC/BAC clones and provides a graphical view of the genome sequence with relevant annotation information at contig level. It also facilitates gene search, statistical analysis of the characteristic features of predicted genes and efficient management of the genome annotation.

## 4.  FUNCTIONAL GENOMICS RESOURCES

The second phase of RGP also focused on elucidating the function of the entire genome. In addition to the existing genome sequencing efforts, several projects on functional analysis of the genome were also launched. These included projects on expression profiling, map-based cloning, molecular breeding, insertional mutagenesis and full-length cDNA analysis. In addition, databases and analyses tools associated with these projects have been constructed and a center for distribution of biological materials generated from these projects has been established.

Analysis of gene function by the gene expression monitoring technique was initiated in 1999 (Yazaki et al., 2000). A system with 1,265 and 8,897 ESTs, respectively, selected from cDNA libraries constructed during the first phase of RGP was established. These rice cDNA microarrays were used to investigate changes in the expression of rice genes at different stages of development and to elucidate genes expressed in response to various types of stress conditions. More than 769 experiments on expression profiling provide the main source of data that can be accessed through the database RED (Rice Expression Database, Yazaki et al., 2002). All normalized expression data can be searched based on the experiment, expression profile and accession of the cDNA clone. Details of the microarray project are also available through RMOS (Rice Microarray Opening Site).

Analyses of agronomically useful genes and complex traits, including QTL, have been effectively accomplished using the precise linkage map with various DNA markers, genomic DNA libraries and physical map. A strategy that involved developing of specific mapping populations and fine mapping of many traits proved effective in map-based cloning. Among the genes that have been isolated with this strategy include *Xa21*, a bacterial blight resistance gene (Yoshimura et al., 1998), *Pib*, a rice blast resistance gene (Wang et al., 1999), a rice gibberellin-insensitive dwarf mutant gene *d*1 (Ashikari et al., 1999) and rice spotted leaf gene *Sp17* (Yamanouchi et al., 2002). Extensive QTL analyses also resulted in characterization of several photoperiod sensitivity genes such as *Hd1* (Yano et al., 2000), *Hd3a* (Kojima et al. 2002) and *Hd6* (Takahashi et al., 2001), as well as genes controlling seed dormancy (Lin et al., 1998) and cool-temperature tolerance at booting stage (Takeuchi et al., 2001).

Insertional mutagenesis is an effective approach in determining the function of 40-60,000 genes predicted in rice. A transposon tagging strategy, which utilizes a rice endogenous retrotransposon *Tos17* has been developed (Hirochika, 2002). Mutations due to *Tos17* transposition are normally induced under tissue culture conditions and are inherited in subsequent generations to facilitate analysis of the mutated gene. So far, more than 50,000 insertional mutant lines carrying about 500,000 insertions have been generated. These resources would be very useful for forward and reverse genetic analyses. The *Tos17* Mutant Panel Database currently contains flanking sequences of *Tos17* insertion sites from 5,000 lines and associated phenotypes. The site also provides a BLAST search against these flanking sequences.

The Rice Full-length CDNA Project has generated sequence data on 175,642 rice full-length cDNAs clustered into 28,469 nonredundant clones (Kikuchi et al., 2003). A total of 21,596 clones have been assigned with tentative protein functions through homology searches of publicly available sequence data. In addition, more than 94% of the clones could be mapped to japonica and indica genomic sequences. All sequence data are available through the database KOME (Knowledge-based

Oryza Molecular biological Encyclopedia). The database provides the nucleotide sequence and encoded amino acid sequence information, results of the homology search with the public databases, mapping information, patterns of alternative splicing, protein domain information, transmembrane structure, cellular localization and Gene Ontology classification. Access to specific information for each full-length cDNA clone can be made by BLAST search, accession number of the clone, specific domain name and general key word search.

These resources are available to the research community through the NIAS DNA Bank and the newly established Rice Genome Resource Center (RGRC). Since 1994, the NIAS DNA Bank has been providing access to biological materials generated mainly from the first phase of RGP including rice ESTs, DNA markers, YAC clones and YAC filters. In addition sequenced PAC and BAC clones are also currently available. The Rice Genome Resource Center was established in April 2003 to provide access to the functional and applied genomics resources derived from the rice genome project. The biological materials currently distributed through this center include full-length cDNA clones, *Tos*17 insertion mutant lines and genetic mapping populations such as recombinant inbred lines (RIL), chromosome segment substitution lines (CSSL) and doubled-haploid lines (DHL) from crosses of different rice varieties. The availability of these resources will allow a wide community of scientists to conduct functional and applied genomics research and to accelerate the application of these resources to improvement of rice and other cereal crops.

## 5. TOWARDS AN INTEGRATED DATA MINING RESOURCE

The major challenge for rice bioinformatics is to establish a comprehensive database that will allow integration of genomic information with present and future expectations in biological and agricultural research. One way to address this need is to interlink the resources of various types of information such as genomic data, phenotypic or expression data, and germplasm resources (Antonio et al., 2000). The Rice PIPELINE which aims to provide a unification tool for rice genomics (Yazaki et al., in preparation) and the Rice Simulator Project which aims to coordinate the data produced from extensive genome analysis with emerging technologies in computational biology should provide the necessary infrastracture for efficient utilization of genomics data.

The Rice PIPELINE integrates the structural and functional genome databases such as KOME, INE, RED, *Tos*17 and PLACE (Higo et al., 1999). A query using a sequence, accession number, clone name or keyword can provide structural information, gene expression information and genome information. The structural information derived through full-length cDNA BLASTN and BLASTX homology provides details of KOME report such as full-length cDNA matches, domain search and GO classification. It also provides a PLACE search for information on the cis-element motif at the 5' upstream region of the full-length cDNA and a *Tos*17 Mutant Panel Database search for phenotype of *Tos*17 flanking sequences. The expression information through the microarray EST BLASTN homology provides gene expression profiles in RED derived from microarray analysis and linked to the genetic and physical mapping information on INE. The genome information through BLASTN with the *japonica* genome or *indica* genome provides links to the sequence data, genetic map and physical map. Thus, any search through the Rice PIPELINE will facilitate integration of genomic information that can be used for elucidating gene structure, identification of function and map-based cloning of useful genes.

The Rice Genome Simulator Project will facilitate integration of the output of genomics research and information technology to build a virtual experimental system designed to estimate the functions of crop genes and make breeding experiments on a computational platform (Harris, 2002). One aspect of the project focuses on the development of a network that will link all fundamental rice genomic resources databases such as the huge amount of data from rice proteome analysis, full-length cDNA, microarray, insertional mutagenesis and biochemical profiling. Another major trust is on the development of an informatics infrastructure including specialized softwares that will allow visualization and integration of various types of complex data, prediction of gene functions and simulation of various parameters. In the long run, the project will focus in simulating rice growth and development by creating in silico models of the rice plant or the so-called "e-rice" at the cellular, tissue and whole organism level. Thus, in the future it would be possible predict how the rice plant will perform under varying environmental or stress conditions. It will also allow breeders to perform crosses between two varieties in silico and select lines with desirable characteristics.

As we move to the postgenomic era over the next ten years, the next challenge for rice bioinformatics is not only in transforming rice genomics data into information that the rice community and biologists in general can query and use properly. The essential logistics necessary to achieve the desired success in rice bioinformatics involve developing the necessary tools for analyzing biological and agricultural problems in multiple dimensions. A robust rice informatics infrastructure that would facilitate integration across various data types may eventually lead to the development of viable strategies for cereal crop improvement.

## 6. REFERENCES

Antonio, B.A., Inoue, T., Kajiya, H., Nagamura, Y., Kurata, N., Minobe, Y., Yano, M., Nakagahra, M., and Sasaki, T., 1996, Comparison of genetic distance and order of genetic markers in five populations of rice, *Genome* **39**:946–956.

Antonio, B.A., Sakata, K., and Sasaki, T., 2000, Rice at the forefront of plant genome informatics, *Genome Informatics* **11**:3–11.

Ashikari, M., Wu, J., Yano, M., Sasaki, T., and Yoshimura, A., 1999, Rice gibberellin-insensitive dwarf mutant gene *Dwarf1* encodes the alpha-subunit of GTP-binding protein, *Proc. Natl. Acad. Sci. USA* **96**:10284–10289.

Baba, T., Katagiri, S., Tanoue, H., Tanaka, R., Chiden, Y., Saji, S., Hamada, M., Nakashima, M., Okamoto, M., Hayashi, M., Yoshiki, S., Karasawa, W., Honda, M., Ichikawa, Y., Arita, K., Ikeno, M., Ohta, T., Umehara, Y., Matsumoto, T., de Jong, P., and Sasaki, T., 2000, Construction and characterization of rice genome libraries: PAC library of *japonica* variety, *Nipponbare* and BAC library of *indica* variety, *NIAR Bull*.**14**:41–49.

Barry, G., 2001, The use of the Monsanto draft rice genome sequence in research, *Plant Physiol.* **125**:1164–1165.

Feng, Q., Zhang, Y., Hao, P., Wang, S., Fu, G., Huang, Y., Li, Y., Zhu, J., Liu, Y., Hu, X., Jia, P., Zhang, Y., Zhao, Q., Ying, K., Yu, S., Tang, Y., Weng, Q., Zhang, L., Lu, Y., Mu, J., Lu, Y., Zhang, L.S., Yu, Z., Fan, D., Liu, X., Lu, T., Li, C., Wu, Y., Sun, T., Lei, H., Li, T., Hu, H., Guan, J., Wu, M., Zhang, R., Zhou, B., Chen, Z., Chen, L., Jin, Z., Wang, R., Yin, H., Cai, Z., Ren, S., Lv, G., Gu, W., Zhu, G., Tu, Y., Jia, J., Zhang, Y., Chen, J., Kang, H., Chen, X., Shao, C., Sun, Y., Hu, Q., Zhang, X., Zhang, W., Wang, L., Ding, C., Sheng, H., Gu, J., Chen, S., Ni, L., Zhu, F., Chen, W., Lan, L., Lai, Y., Cheng, Z., Gu, M., Jiang, J., Li, J., Hong, G., Xue, Y., and Han, B., 2002, Sequence and analysis of rice chromosome 4, *Nature* **420**:316–320.

Goff, S.A., Ricke, D., Lan, T.H., Presting, G., Wang, R., Dunn, M., Glazebrook, J., Sessions, A., Oeller, P., Varma, H., Hadley, D., Hutchison, D., Martin, C., Katagiri, F., Lange, B.M., Moughamer, T., Xia, Y., Budworth, P., Zhong, J., Miguel, T., Paszkowski, U., Zhang, S., Colbert, M., Sun, W., Chen, L., Cooper, B., Park, S., Wood, T.C., Mao, L., Quail, P., Wing, R., Dean, R., Yu, Y., Zharkikh, A., Shen, R., Sahasrabudhe, S., Thomas, A., Cannings, R. Gutin, A., Pruss, D., Reid, J., Tavtigian, S., Mitchell, J., Eldredge, G., Scholl, T., Miller, R.M., Bhatnagar, S. Adey, N., Rubano, T., Tusneem, N., Robinson, R., Feldhaus, J. Macalma, T., Oliphant, A., and Briggs, S., 2002, A draft sequence of the rice genome (*Oryza sativa* L. ssp. *japonica*), *Science* **296**:92–100.

Harris, S.B., 2002, Virtual rice, *EMBO Reports* **3**:511–513.

Harushima, Y., Yano, M., Shomura, A., Sato, M., Shimano, T., Kuboki, Y., Yamamoto, T., Lin, S.Y., Antonio, B.A., Parco, A., Kajiya, H., Huang, N., Yamamoto, K., Nagamura, Y., Kurata, N., Khush, G.S., and Sasaki, T., 1998, A high-density rice genetic linkage map with 2,275 markers using a single F2 population, *Genetics* **148**:479–494.

Hirochika, H., 2001, Contribution of the *Tos*17 retrotransposon to rice functional genoics, *Curr. Opin. Plant Biol.* **4**:118–122.

Higo, K., Ugawa, Y., Iwamoto, M., and Korenaga, T., 1999, Plant cis-acting regulatory DNA elements (PLACE) database, *Nucleic Acids Res.***27**:297–300.

Kikuchi, S., Satouh, K., Nagata, T., Kawagashira, N., Doi, K., Kishimoto, N., Yazaki, J., Ishikawa, M., Yamada, H., Ooka, H., Hotta, I., Kojima, K., Namiki, T., Ohneda, E., Yahagi, W., Suzuki, K., Li, C.J., Ohtsuki, K., Shishiki, T., Otomo, Y., Murakami, K., Iida, Y., Sugano, S., Fujimura, T., Suzuki, Y., Tsunoda, Y., Kurosaki, T., Kodama, T., Masuda, H., Kobayashi, M., Xie, Q., Lu, M., Narikawa, R., Sugiyama, A., Mizuno, K., Yokomizo, S., Niikura, J., Ikeda, R., Ishibiki, J., Kawamata, M., Akemi, Y., Miura, J., Kusumegi, T., Oka, M., Ryu, R., Ueda, M., Matsubara, K., Kawai, J., Carninci, P., Adachi, J., Aizawa, K., arakawa, T., Fukuda, S., Hara. A., Hashidume, W., Hayatsu, N., Imotani, K., Ishii, Y., Itoh, M., Kagawa, I., Kondo, S., Konno, H., Miyazaki, A., Osato, N., Ota, Y., Saito, R., Sasaki, D., Sato, K., Shibata, K. Shinagawa, A., Shiraki, T., Yoshino, M., and Hayashizaki, Y., 2003, Collection, mapping, and annotation of over 28,000 cDNA clones from japonica rice, *Science* **301**:376–379.

Kojima, S., Takahashi, Y., Kobayashi, Y., Monna, L., Sasaki, T., Araki, T., and Yano, M. 2002, *Hd3a*, a rice ortholog of the Arabidopsis*FT* gene, promotes transition to flowering downstream of *Hd1* under short-day condition, *Plant Cell Physiol.* **43**:1096–1105.

Kurata, N., Nagamura, Y., Yamamoto, K., Harushima, Y., Sue, N., Wu, J., Antonio, B.A., Shomura, A,. Shimizu, T., Lin, S.Y., Fukuda, A., Shimano, T., Kuboki, Y., Toyama, T., Miyamoto, Y., Kirihara, T., Hayasaka, K., Miyao, A., Monna, L., Zhong, H.S., Tamura, Y., Wang, Z.X., Momma, T., Umehara, Y., Yano, M., Sasaki, T., and Minobe, Y., 1994b, A 300 kilobase interval genetic map of rice including 883 expressed sequences, *Nature Genetics* **8**:365–372.

Kurata, N., Moore, G., Nagamura, Y., Foote, T., Yano, M., Minobe, Y., and Gale, M., 1994a, Conservation of genome structure between rice and wheat, *Bio/Technology* **12**:276–278.

Lin, S.Y., Sasaki, T., and Yano, M., 1998, Mapping quantitative trait loci controlling seed dormancy and heading date using backcross inbred lines in rice, *Oryza sativa* L., *Theor. Appl. Genet.* **96**:997–1003.

Moore, G., Devos, K.M., Wang, Z., and Gale, M.D., 1995, Grasses, line up and form a circle, *Current Biol.* **5**: 737–739.

Saji, S., Umehara, Y., Antonio, B.A., Yamane, H., Tanoue, H., Baba, T., Aoki, H., Ishige, N., Wu, J., Koike, K., Matsumoto, T., and Sasaki, T., 2001, A physical map with yeast artificial chromosome (YAC) clones covering 63% of the 12 rice chromosomes, *Genome* **44**:32–37.

Sakata, K., Nagamura, Y., Numa, H., Antonio, B.A., Nagasaki, H., Idonuma, A., Watanabe, W., Shimizu, Y., Horiuchi, I., Matsumoto, T., Sasaki, T., and Higo, K., 2002, Rice GAAS: an automated annotation system and database for rice genome sequence, *Nucleic Acids Res.* **30**:98–102.

Sakata, K., Antonio, B.A., Mukai, Y., Nagasaki, H., Sakai, Y., Makino, K., and Sasaki, T., 2000, INE: a rice genome database with an integrated map view, *Nucleic Acids Res.* **28**:97–101.

Sasaki, T., Matsumoto, T., Yamamoto, Y., Sakata, K., Baba, T., Katayose, Y., Wu, J., Niimura, Y., Cheng, Z., Nagamura, Y., Antonio, B.A., Kanamori, H., Hosokawa, S., Masukawa, M., Arikawa, K., Chiden, Y., Hayashi, M., Okamoto, M., Ando, T., Aoki, H., Arita, K., Hamada, M., Harada, C., Hijishita, S., Honda, M., Ichikawa, Y., Idonuma, A., Iijima, M., Ikeda, M., Ikeno, M., Ito, S., Ito, T., Ito, Y., Ito, Y., Iwabuchi, A., Kamiya, K., Karasawa, W., Katagiri, S., Kikuta, A., Kobayashi, N., Kono, I., Machita, K., Maehara, T., Mizuno, H., Mizubayashi, T., Mukai, Y., Nagasaki, H., Nakashima, M., Nakama, Y., Nakamichi, Y., Nakamura, M., Namiki, N., Negishi, M., Ohta, I., Ono, N., Saji, S., Sakai, K., Shibata, M., Shimokawa, T., Shomura, A., Song, J., Takazaki, Y., Terasawa, K., Tsuji, K., Waki, K., Yamagata, H., Yamane, H., Yoshiki, S., Yoshihara, R., Yukawa, K., Zhong, H., Iwama, H., Endo,T., Ito, H., Hahn, J.H., Kim, H.I., Eun, M.Y., Yano, M., Jiming, J., and Gojobori, T., 2002, The genome sequence and structure of rice chromosome 1, *Nature* **420**:312–316.

Sasaki, T., and Burr, B., 2000, International rice genome sequencing project: the effort to completely sequence the rice genome, *Curr. Opin. Plant Biol.* **3**:138–141.

Sasaki, T., 1999, Rice genome research in Japan as a frontrunner in crop genomics. *Sci Technol. Jpn.* **18**:18–21.

Takahashi, Y., Shomura, A., Sasaki, T., and Yano, M., 2001, *Hd6*, a rice quantitative trait locus involved in photoperiod sensitivity, encodes the a subunit of protein kinase CK2, *Proc. Natl. Acad. Sci. USA* **98**:7922–7927.

Takeuchi, Y., Hayasaka, H., Chiba, B., Tanaka, I., Shimano, T., Yamagishi, M., Nagano, K., Sasaki, S., and Yano, M., 2001, Mapping quantitative trait loci controlling cool-temperature tolerance at booting stage using doubled-haploid lines of temperate japonica rice cultivars, *Breed. Sci.* **51**:191–197.

Wang, Z.X., Yano, M., Yamanouchi, U., Iwamoto, M., Monna, L., Hayasaka, H., Katayose, Y., and Sasaki, T., 1999, The *Pib* gene for rice blast resistance belongs to the nucleotide binding and leucine-rich repeat class of plant disease resistance genes, *Plant J.* **19**:55–64.

Wu, J., Maehara, T., Shimokawa, T., Yamamoto, S., Harada, C., Takazaki, Y., Ono, N., Mukai, Y., Koike, K., Yazaki, J., Fujii, F., Shomura, A., Ando, T., Kono, I., Waki, K., Yamamoto, K., Yano, M., Matsumoto, T., and Sasaki, T., 2002, A comprehensive rice transcript map containing 6591 expressed sequence tag sites, *Plant Cell* **14**:525–535.

Yamamoto, K., and Sasaki, T., 1997, Large-scale EST sequencing in rice. *Plant Mol. Biol.* **35**:135–144.

Yano, M., and Sasaki, T., 1997, genetic and molecular dissection of quantitative traits in rice, *Plant Mol. Biol.* **35**:145–153.

Yano, M., Katayose, Y., Ashikari, M., Yamanouchi, U., Monna, L., Fuse, T., Baba, T., Yamamoto, K., Umehara, Y., Nagamura, Y., and Sasaki, T., 2000, *Hd1*, a major photoperiod sensitivity quantitative trait locus in rice, is closely related to the Arabidopsis flowering time gene *CONSTANS*, *Plant Cell* **12**:2473–248.

Yazaki, J., Kishimoto, N., Nakamura, K., Fujii, F., Shimbo, K., Otsuka, Y., Wu, J., Yamamoto, K., Sakata, K., Sasaki, T., and Kikuchi, S., 2000, Embarking on rice functional genomics via cDNA microarray: Use of 3′ UTR probes for specific gene expression analysis, *DNA Res.* **7**:367–370.

Yazaki, J., Kishimoto, N., Ishikawa, M., and Kikuchi, S., 2002, Rice Expression Database: the gateway to rice functional genomics, *Trends Plant Sci.* **7**:563–564.

Yazaki, J., Kojima, K., Suzuki, K., Kishimoto, N., and Kikuchi, S., The Rice PIPELINE: a unification tool for plant functional genomics, (in preparation).

Yoshimura, S., Yamanouchi, U., Katayose, Y., Toki, S., Wang, Z.X., Kono, I., Kurata, N., Yano, M., Iwata, N., and Sasaki, T., 1998, Expression of *Xa1*, a bacterial blight-resistance gene in rice, is induced by bacterial inoculation, *Proc. Natl. Acad. Sci. USA* **95**:1663–1668.

*Chapter 2*

# Application of Evolutionary Computation to Bioinformatics

Daniel Ashlock

## 1. INTRODUCTION

In solving a scientific problem, one of the most helpful possibilities is that you will see a pattern in your data. It is almost the definition of an interesting scientific problem that it contains some sort of pattern. The patterns that arise in nature are often subtle and escape notice until cleverness or hard work un-cover them. The field of machine learning is a collection of techniques intended to automate the process of pattern discovery. A broad survey of machine learning techniques applied to bioinformatics is given in Pierre Baldi and Soren Brunak (2001). This document introduces a single, relatively versatile machine learning technique called evolutionary computation. A collection of applications of evolutionary computation to bioinformatics is given in Fogel and Corne (2003).

Both machine learning and evolutionary computation have applications far beyond bioinformatics but almost all of the techniques in the domain of machine learning and evolutionary computation have useful applications within bioinformatics. The introduction to evolutionary computation given here is in the form of three examples intended to showcase three substantially different applications of evolutionary computation. The first, while it solves a real problem, is an almost trivial instance of evolutionary computation. It seeks a gapless alignment of 315 sequences in a fashion that permits the discovery of a motif associated with

**Daniel Ashlock**     Department of Mathematics, Bioinformatics and Computational Biology Program, Iowa State University, Ames, IA 50010

*Genome Exploitation: Data Mining the Genome*, edited by J. Perry Gustafson, Randy Shoemaker, and John W. Snape.
Springer Science + Business Media, New York, 2005.

**Table 1**
**Predictions versus Truth Results for the Most Fit**
**Finite State Machines Located During the First Set**
**of Evolutionary Runs.**

| | *Training Data* | | | | *Crossvalidation Set* | | |
|------|-----|-----|-----|------|-----|-----|-----|
| | Prediction | | | | Prediction | | |
| | + | − | ? | | + | − | ? |
| Good | 666 | 256 | 78 | Good | 115 | 106 | 29 |
| Bad | 287 | 659 | 54 | Bad | 96 | 125 | 29 |

insertion of a mu-transposon. The second example is a more sophisticated but standard application of evolutionary computation, learning patterns in a collection of good and bad primers designed as part of a *Zea mays* genomics project. Once learned, these patterns are then used to reduce the failure rate of primers in subsequent work. The third example is a departure from standard evolutionary algorithms, which fuses evolutionary algorithms with greedy algorithms to create a new type of evolutionary algorithm called a greedy closure genetic algorithm. This algorithm is used to create error correcting DNA bar codes for use in pooled genetic libraries. These bar codes permit the identification of the library, which contributed a given expressed sequence tag. The error correcting property of the bar codes permits the identification of the library even when the area containing the bar code is sequences with some errors.

## 2. EVOLUTIONARY COMPUTATION

Evolutionary computation has been described as a "Swiss army algorithm" in comparison to the Swiss Army knife, which typically has a whole collection of small tools built into it (Fig. 1). This description is both misleading and a good starting point for discussing the strengths of evolutionary computation. A completely general-purpose tool would be useless (probably too heavy to lift). The Swiss army knife is not a multipurpose tool. Careful examination will show the alert scientist that it is a collection of many tools with fairly specific applications. Sure, it is possible to use the screwdriver as a hole punch and the large blade may be useful for both shaving a dowel and trimming a bad spot out of an apple, but the applications of each tool within the knife are fairly limited. Also, the tools you are not using tend to get in your way. The purchase on a Swiss army knife is not as good as the handle on a normal screwdriver. What does this have to do with evolutionary computation? The basic algorithm for evolutionary computation is given in Table 1. This algorithm can solve any problem for which the solutions can be placed in the form of a data structure and for which the quality of the solutions in such data structures can be compared. In spite of this apparent power, the process of getting a problem transformed into a data structure and then creating a useful quality comparison can be insuperably difficult. Even if these hurdles can be overcome, the running time on evolutionary algorithms is typically long and so

```
Generate an initial population of solutions.
Repeat
        Evaluate your solutions.
        With a quality bias, select solutions.
        Reproduce and vary the solutions selected.
        Place the new solutions in the population.
Until(Satisfied)
```

**FIGURE 1.** Basic algorithm for evolutionary computation.

an evolutionary algorithm may be impractical. Let us start with the very simplest type of evolutionary algorithm, a string evolver.

In a string evolver the population is a population of character strings and the notion of fitness is that of matching a reference string. Examine the following population member (character string), aligned with the reference string "Madam, I'm Adam.":

**Reference: "Madam, I'm Adam."**

**Population Member: "Mad*g,hI.m Admm!"**

**Fitness loci: +++ + + ++++ +**

The fitness of the above population member is 10 of a possible 16 because 10 of its 16 characters match the reference string. Once we have a notion of fitness it becomes possible to apply the algorithm given in Table 1.

Generating an initial population of character strings is done by filling in characters at random in each member of the population. We pick strings to reproduce by shuffling the population into groups of four and permitting the best two members of each group of four to reproduce. Their offspring replace the two worst members of the group of four. The relatively small size of the groups (four members) represents a weak bias in favor of fitness; larger groups would yield sharper selection. Reproduction is done by first copying the two strings that are reproducing and then performing *crossover* and *mutation* on the copies. These words have completely different meanings in the context of evolutionary algorithms than they do in biology. Crossover consists of exchanging middle segments of the strings; mutation consists of picking a position in the string and putting in a new random character. An example of crossover is shown in Figure 2, an example of mutation is shown in Figure 3.

The process of evolving a copy of a string that you already have in hand is not an intrinsically interesting one. It does serve as a simple example of evolutionary computation and serves as a starting point for discussing the design of evolutionary

```
                                          Fitness:
          Parents: "Mad*g,hI.m Admm!"       10
                   "kadam, I'm Adasl"        13

          Children: "Madam, I'm Adam!"       15
                    "kad*g,hI.m Admsl"        8
```

**FIGURE 2.** Depicted above is crossover. Two selected parent strings are copied and characters in positions 3–14 are swapped. We call 3 and 14 the *crossover points* and the type of crossover depicted is called *two point crossover*.

```
            Child: "Madam, I'm Adam!"   15
                         |
        Mutated Child: "Ma^am, I'm Adam!"   14
```

**FIGURE 3.** Mutation consists of taking one or more character positions and generating new random characters at those positions. Above is depicted mutation at the third character in a string with a net decrease in fitness.

algorithms. A trace of the best string found so far in a run of a string evolver is shown in Figure 4. Notice that the time to the next improvement in fitness is highly irregular but generally increases with time. Initially, crossover can bring together correct sub-strings. As evolution proceeds the population in the algorithm becomes highly inbred and crossover does little. Toward the end of a run, mutation becomes the sole source of progress as we wait for a fortuitous mutation to fill in the one or two missing characters not present in the initial population.

The method of picking parents and placing offspring used in the string evolver example is called *single tournament selection with size four.* The methods of producing variation are called *two point crossover and single point mutation.* There is a wealth of possible choices for parent selection, offspring placement, and of variation producing techniques. Many of these are detailed in Ashlock (2004), which is available at *http://www.math.iastate.edu/danwell/EC*

## 3. FINDING A TRANSPOSON INSERTION MOTIF

In Dietrich, Cui, Packila, Ashlock, Nikolau, and Schnable, (2002), an example of an application of a string evolver to bioinformatics appears. A collection of 315

| | | Appeared in |
|---|---|---|
| Best String | Fitness | Generation |
| HadDe Q'/--<jlm' | 3 | 5 |
| HadDe,em3m/<I jm- | 4 | 52 |
| HadDe,em3m/<I jm- | 5 | 54 |
| HadDm,ex3m/#I jmj | 6 | 73 |
| HadDm,eI8m/#I jmj | 7 | 86 |
| HadDm,eI8m[Aj jmt | 8 | 118 |
| HadDm,UI8m[Aj jm. | 9 | 135 |
| MadDm,zI8m4AJ1m. | 10 | 154 |
| Madam,zIXm4AJ1m. | 11 | 163 |
| Madam, InmqAJym. | 12 | 256 |
| Madam, I'mqArHm. | 13 | 327 |
| Madam, I'm AC⁻m. | 14 | 473 |
| Madam, I'm APam. | 15 | 512 |
| Madam, I'm Adam. | 16 | 647 |

**FIGURE 4.** A string evolver running. Each time an improvement in the best fitness in the population occurs, the string receiving that fitness is printed together with the generation number in which it appeared.

DNA sequences 129 bases long and centered on a 9 base repeat created during a transposon insertion were acquired. The conjecture is that there is a motif favored by the transposon at the site of insertion but this motif is masked by not knowing which orientation of the sequenced DNA contains the motif. The motif is not tight enough to permit its discovery by inspection or simple statistical analysis. What is desired is to recover the correct orientation by aligning all 315 sequences. There is not sufficient sequence homology at the insertion sites to permit useful alignment via dynamic programming, the standard alignment technology.

The choice of data structure for attempting to solve this problem is fairly obvious. It is a character string of 315 zeros and ones that specifies an alignment of the sequences. A zero means "leave the sequence in it current orientation" while a 1 means "compute the reverse complement of the sequence." This yields a space of 2314 = 3:34 × 1094 possible alignments (the first sequence is never reversed, selecting one of two possible global alignments). The tricky part of designing the evolutionary algorithm in this case is the fitness function. It was decided to minimize the randomness of the alignment of the sequences. In this case the randomness of an alignment (choice of forward or reverse orientations) was estimated by computing the squared deviation of the number of bases of each type at each position from the empirical frequency of those bases in the sequences being aligned. This measure of non-randomness is given in Equation 1.

$$f(A) = \sum_{i=1}^{129} \left( \sum_{z \in (C,G,A,T)} (N_x - E_x)^2 \right)$$

The number $N_x$ is the number of bases x at a given position while $E_x$ is the expected number of bases of that type, i.e. the number of sequences times the fraction of all bases in the data set of type x.

Using Equation 1 as a fitness function, a string evolver can be used to perform the alignment. The string is made up of characters each of which specifies the orientation of one of 314 sequences with the first sequence left in its original orientation. Such a string evolver that searched the space of alignments was run 100 times. Out of those 100 runs, 88 returned the same alignment and the same fitness value. This fitness value was the largest found in any of the runs. This makes it likely we are detecting a true optimal alignment of the 315 sequences.

In order to see if there is a motif at or near the point of insertion, a $x^2$-value was computed for each position of the alignment. The $x^2$-values are given along the length of the alignment in Figure 5. The unaligned sequences did not show a significantly non-random base composition at any point. The aligned sequences yield a significant deviation from the expected base composition statistics at points immediately flanking the insertion. Readers interested in the biology that underlies this example should read Dietrich, Cui, Packila, Ashlock, Nikolau, and Schnable, (2002).

It is important to consider the question of interaction between the fitness function and the $x^2$-statistics. In this case the non-randomness of the entire alignment was maximized. Using 60 bases of flanking sequence on either side of the insertion

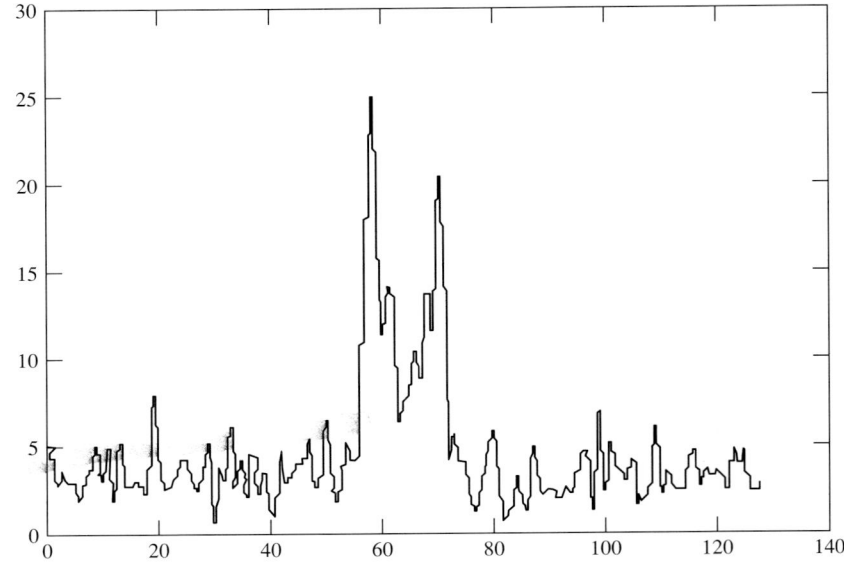

**FIGURE 5.** $x^2$-values derived from base composition at each position in an alignment of 315 DNA sequences of length 129 fanking distinct insertion points of a mu-transposon. The alignment used was the best found by the evolutionary algorithm.

point reduces the chance that we are creating a motif by fortuitous arrangement of existing variation in a small region including the point of insertion. The fact that the $x^2$-values spike in positions immediately flanking the insertion but not in the remainder of the flanking sequence suggests a motif does exist at the point of insertion. It is also important to note that non-randomness was maximized relative to the empirical base statistics of the sequences aligned. Using some larger region to compute expected base frequencies would have made it easier to create a phantom motif by exploiting and aligning existing random sequence features.

## 4. PCR PRIMER PICKING

A more complex application of evolutionary computation to bioinformatics is that of picking PCR primers. At this point the application is complex enough to raise the issue of representation. The representation used in an instance of evolutionary computation is the way that candidate solutions to a problem are coded as data structures and varied during reproduction. The representation used to align the transposon insertion sites was a character string representation with two point crossover and single point mutation, exactly the same representation as was used in the example string evolver. Before picking a representation for primer picking, we will need a clear specification of the problem.

Using a standard primer-picking tool (Primer 3 from the NCSA Biology Workbench), many thousands of primers were designed to amplify sites likely to contain polymorphisms in *Zea mays*. Many of these primers amplified their targets correctly, while others did not. The problem is to distinguish the good primers from the bad primers given that the original primer picking software thought they were all good. We are neglecting technician error in performing the PCR reactions and in scoring the outcomes of the PCR experiments, in effect treating these sources of errors as "noise". We thus act as if the scoring of primers as good or bad is entirely correct. These primers, scored as good and bad, form the *training data* for our primer picking system.

The experiment designed to find bad primers used evolutionary computation as a machine learning system to attempt to detect any patterns that will help us to tell good primers from bad primers. Note that many of the standard things that make a primer good, such as correct Tm and the presence of a GC-clamp are already in every primer because they were put there by the original primer picking software. This means we will be looking for organism specific patterns that only affect primers in *Zea mays*. In subsequent work on *Zea mays*, multiple primers will be designed for each target and the good/bad classifier created via machine learning will pick from among them those primers most likely to work based on patterns learned from earlier primers.

In the last example there was an obvious representation (a character string that specified sequence orientations) but the fitness function (minimize overall randomness of the alignment) was not such a clear choice. In building an evolutionary algorithm to pick primers, it turns out that both the fitness function and representation are not obvious. We want a classifier that, given a primer, gives a (often correct) prediction if it is a good or bad primer. Fitness will thus consist of some abstraction of predicting correctly the good/bad status of the primers in the training data. In a set of initial experiments simply scoring the number of correct predictions did not work well.

Given that the problem could be in a motif that is somewhere in the primer, we need a representation that does not assess bases in a manner based on their distance from the end of the primer but rather based on the pattern of surrounding bases. Because of this need to have a non-position-specific assessment we settled on finite state machines as our representation. A finite state machine can process a string of bases, waiting for one of a small set of motifs to appear, and then make a state transition that detects it. The widely used BLAST software incorporates finite state machines to perform essentially this task. As an additional benefit, use of finite state machines permitted a unique sort of incremental fitness reward, described subsequently.

A finite state machine is a collection of states (including a starting state), together with a collection of data-driven transitions among the states. The output of a finite state machine is associated with the transitions or the states themselves. An example of a finite state machine is shown in Figure 6. In this case the output of the FSM is encoded by noting that, when it is in state five, the last three bases it encountered formed a stop codon. In order to give an incremental assessment of

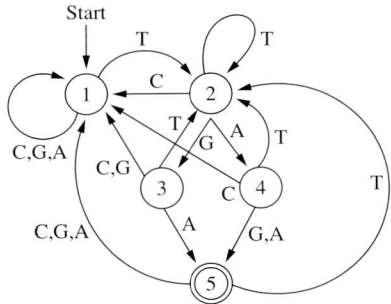

**FIGURE 6.** A finite state machine that recognizes stop sequences in DNA.

strings of DNA bases we are considering as primers we will make a modification to this standard sort of finite state machine.

The states of the finite state machines used to classify primers have three possible types or labels: ? (don't know), + (good primer), and −(bad primer). These state labels are used to permit the finite state machine to function as a classifier. The fitness of a finite state machine on a training set of primers is computed as follows. Each PCR primer in a set of training data is run through the finite state machine. As the machine passes through each state it is given +1 score if the state label matches the good/bad status of the primer and −1 if it doesn't match. No incremental score is awarded for the don't know states. Fitness is summed over all primers examined. If we imagine the evolutionary algorithm as searching a fitness landscape for good classifiers then the use of this sort of incremental reward scheme acts to smooth the landscape and permit the search to avoid getting stuck. A more complete discussion of these issues appears in Ashlock, Wittrock, and Wen, (2002). A finite state machine of the sort used to classify primers is shown in Figure 7 as both a state transition diagram (picture) and as a table.

Having decided to use finite state machines in our evolutionary computation system, we still need to select methods for generation of an initial population and for generating variations during reproduction. The finite state machines are initialized uniformly at random, filling in both transitions and state labels with uniformly distributed valid values. As with the string evolver, we will have a crossover operator and a kind of mutation. Both of these variation operators need to be re-tooled to contend with the more complex structure of finite state machines. The crossover operator used works by treating the states, including their label and outward transitions, as "characters" and then performing crossover on the "string" of states. Two point crossover of this string of states was used and the designation of the initial state moves with the first state during crossover. The mutation used on the finite state machine modifies the choice of initial state 10% of the time, randomly picks a new destination for one of the transitions 30% of the time, and modifies the label f+;−; ?g on a state 60% of the time.

One hundred evolutionary runs with distinct random starting populations of 600 machines were performed. The best finite state machine from each simulation

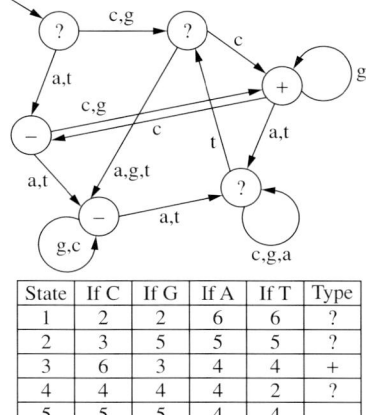

| State | If C | If G | If A | If T | Type |
|-------|------|------|------|------|------|
| 1 | 2 | 2 | 6 | 6 | ? |
| 2 | 3 | 5 | 5 | 5 | ? |
| 3 | 6 | 3 | 4 | 4 | + |
| 4 | 4 | 4 | 4 | 2 | ? |
| 5 | 5 | 5 | 4 | 4 | − |
| 6 | 3 | 3 | 5 | 5 | − |

**FIGURE 7.** A finite state machine configured for primer classification. Both a pictorial and a tabular representation are given. The starting state, denoted by the rootless arrow, is state 1 and the states in the pictorial representation are numbered clockwise from that point.

was saved for evaluation as a classifier and for use in later sets of runs. Each evolutionary run proceeded for 1000 generations. These evolutionary runs used a set of 2000 primers, half good and half bad, as their training data. In order to ensure that the classifers were learning patterns rather than just memorizing the training data they were cross-validated on a set of 500 primers, also half good and half bad, that were not part of the training data. In order to use the finite state machines to predict the good/bad status of a primer, primers are run through the finite state machines noting how many of each type of state are encountered. A majority vote is taken on the type of state label encountered. This permits a failure to classify if a majority of the states are of type ?. Table 1 documents the classification abilities of the best finite state machine located in the first 100 runs.

The outcomes given here show some patterns are being located but the finite state machines are not yet classifying well enough to have a substantial impact on the number of bad primers used. A score of 240 correct, 192 wrong, and 59 undecided is less than one would hope for. Examining the distribution of best finesses it appeared that a few of the evolutionary runs had discovered interesting patterns and many had not. In an attempt to consolidate these patterns in a single finite state machine we performed a second set of evolutionary runs that hybridized the finite state machines found in the first set of runs. For another instance of hybridization in the context of evolutionary computation see (Ashlock and Joenks, 1998). The hybridization runs are identical to the first set except that 100 members of the initial random population are replaced with the best-of-run finite state machines saved during the first 100 runs. The other members of these initial populations are still generated uniformly at random. A second set of hybridizations was performed,

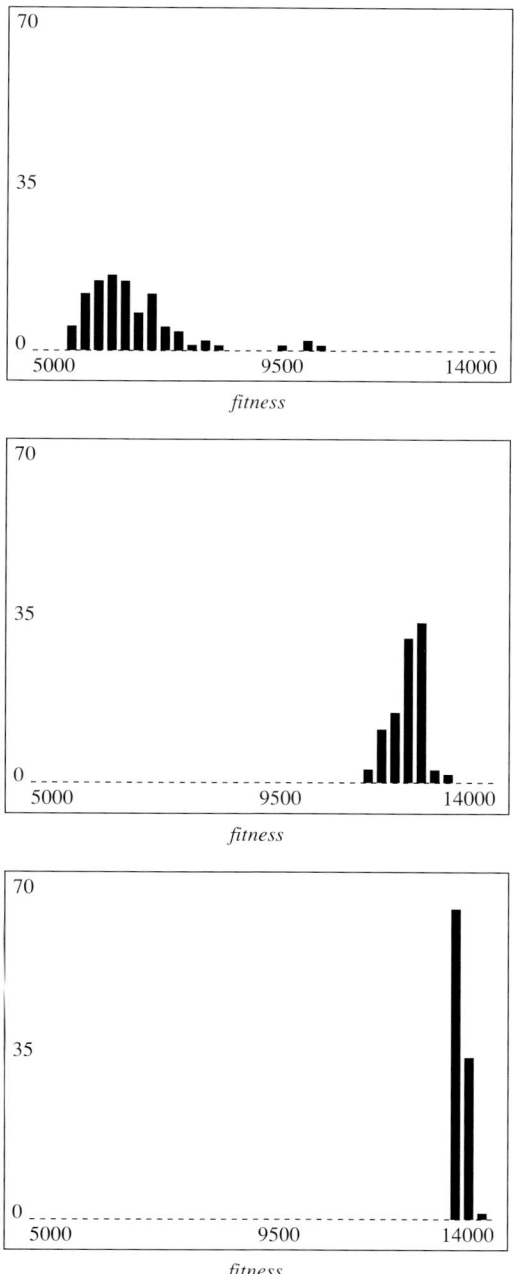

**FIGURE 8.** Histograms of the distribution of fitnesses. Top to bottom, for the first set of simulations, for the first set of hybridizations, and for the second set of hybridizations.

enough to have an impact on primer location. It makes 265 correct predictions, 216 bad predictions, and cannot decide about 19 of the primers. Since different runs may have found classifiers that recognize different patterns, it is also possible that a voting scheme among several classifiers will yield even better performance. That possibility has not yet been evaluated.

## 5. DNA BAR CODE LOCATION

Many useful evolutionary algorithms are hybrids. The word "hybrid" is used in a completely different fashion from its use in the preceding section. In this case a hybrid algorithm is a combination of evolutionary computation with some other type of algorithm. In this case we combine evolutionary computation with a *greedy algorithm*. We will use this greedy algorithm, under the control of evolved structures, to locate an error correcting bar code over the DNA alphabet. The errors corrected will not be biological but rather will permit us to recognize embedded DNA bar codes in constructs even when a sequencing error has modified them.

Greedy algorithms are familiar to most computational professionals. They are algorithms that use a greedy rule (e.g. make the best possible next move) to try to accomplish some goal. A few greedy algorithms, like those for finding a minimal-weight spanning tree in a weighted network, can be proven to yield optimal results. Other problems, like graph 14 coloring or the traveling salesman problem, admit a plethora of greedy algorithms all of which cannot be shown to yield optimal results. While it would seem that the control and improvement of greedy algorithms is a natural target for evolutionary computation, relatively few attempts have been made. It turns out that there are several possible approaches. The approach explored in the current paper consists of making small modifications in the order of presentation of potential parts of a growing structure as a means of deflecting the greedy algorithm's behavior. The role of evolutionary computation is to locate modifications of the order of presentation that result in better structures. The resulting structure is not a standard one for evolutionary computation and the crossover used is probably macromutational (Peter, 1997) in character. The greedy algorithm we will use is Conway's lexicode algorithm.

### Algorithm 1 Conway's Lexicode Algorithm
*Input: A minimum distance d and a word length n.*
*Output: An(n,d) − code.*
*Algorithm:*

*Place the DNA words of length n in lexicographical order. Initialize an empty set C of words. Scanning the ordered collection of DNA words, select a word and place it in the code if it is distance d or more from each word placed in code so far.*

An (n; d)-error correcting code is a collection of n-character strings (in this case strings of DNA) that have the property that any two of them are at least d errors apart. The notion of error must be chosen to fit the problem. In this case, since sequencing errors can change, remove, or insert an apparent DNA base, the relevant notion of error is the edit metric. In the edit metric the distance between two strings of DNA is the minimum number of single base additions, deletions, or substitutions required to transform one string into the other. The edit distance of CGATT and GGAT, for example, is two: change the initial C to a G and delete the terminal T. Since no single edit will transform one of these strings into the other, their edit distance is exactly 2.

The (n; d)-error correcting codes over the DNA alphabet will be used as error tolerant bar codes for genetic constructs. In order to correct errors, we note that any collection of less that 1/2d errors leaves us nearer to the actual bar code used than any other. Error correction consists of checking what bar code, among those used in any genetic construct, is closest in edit distance to the one apparently sequenced at the bar code's location in the genetic construct.

Examining Algorithm 1, the reader will see that the algorithm extends a partial code as it goes along. Since the algorithm considers the potential code words in a fixed order and starts with an empty set of code words the algorithm is deterministic. Run many times it always produces the same result. In order to get a different code out of the algorithm we would need to present the words to the algorithm in a different order. A *greedy closure evolutionary algorithm* exploits a restricted method of permuting the order in which the words are considered. Conway's lexicode algorithm is modified by specifying a short list of words, called a seed, that start as members of the code. The seeds used here have three members. Once a seed is chosen, Conway's algorithm is used to "close" the partial code represented by the seed. The words not in the seed are presented in the same order as in the standard algorithm. Since each word chosen to be in the code prevents any other word within d edits from being in the code even a small seed can have a huge impact on the membership and size of the code.

The structure to be evolved by the evolutionary algorithm is the seed. The fitness of a seed is the size of the resulting error correcting code. Notice that bigger codes are better because you get more bar codes with the same error correcting potential.

We have chosen seeds as the data structure to evolve. In order to complete a representation for the bar code location problem we must also specify the crossover and mutation operators for the data structure. Crossover of two seeds consists of copying two parents and then randomly shuffling the words between the copies, save that any words that appear twice send one copy to each child. Mutation consists of replacing one word in a seed with a new word generated at random. All of the variation operators listed here have the potential to create seeds, which violate the minimum distance for the code. Such seeds are awarded a fitness of zero and so removed from the population by the selection process. We call these seeds invalid. For valid seeds, the size of the code resulting from application of the lexicode algorithm to the seed is the fitness of that seed.

Before we run an evolutionary algorithm to locate bigger error correcting codes (equivalently: sets of DNA bar codes) it would be a good idea to compute

**Table 4**

**Size of DNA Edit-Metric Lexicodes Found with the Unmodified Version of Conway's Lexicode Algorithm.**

| Code Size | Minimum Distance $d$ | | | | | | |
|---|---|---|---|---|---|---|---|
| Length n | 3 | 4 | 5 | 6 | 7 | 8 | 9 |
| 3 | 4 | – | – | – | – | – | – |
| 4 | 12 | 4 | – | – | – | – | – |
| 5 | 36 | 8 | 4 | – | – | – | – |
| 6 | 96 | 20 | 4 | 4 | – | – | – |
| 7 | 311 | 57 | 14 | 4 | 4 | – | – |
| 8 | 1025 | 164 | 34 | 12 | 4 | 4 | – |
| 9 | 3451 | 481 | 90 | 25 | 10 | 4 | 4 |
| 10 | * | 1463 | 242 | 57 | 17 | 9 | 4 |
| 11 | * | * | 668 | 133 | 38 | 13 | 4 |

\* Big
- empty

the rough size we could expect codes to be. Table 4 gives the sizes of the codes located by the unmodified version of Conway's algorithm for several values of the length n of the bar code and the minimum distance d between any two strings in the code.

The evolutionary algorithm was run 100 times for 100 generations in for each of ten different sets of parameters n and d. Results from these runs are given in Table 5. The greedy fitness evolutionary algorithm outperformed the plain lexicode algorithm for all parameter sets tested. An example of a (6; 3)-error correcting code in DNA for the edit metric is given in Table 6.

**Table 5**

**Comparison of DNA Edit Metric Code Sizes for the Plain Lexicode Algorithm and the Greedy Fitness Evolutionary Algorithm. The Figures in Parenthesis are the Number of Times the Best Result was Located in 100 Runs.**

| Length | Minimum Distance | Plain Lexicode | Evolutionary Algorithm |
|---|---|---|---|
| 4 | 3 | 12 | 16 (18) |
| 5 | 3 | 36 | 41 (2) |
| 5 | 4 | 8 | 11 (1) |
| 6 | 3 | 96 | 106 (2) |
| 6 | 4 | 20 | 25 (11) |
| 6 | 5 | 4 | 9 (9) |
| 7 | 3 | 311 | 329 (2) |
| 7 | 4 | 57 | 63 (1) |
| 7 | 5 | 14 | 18 (12) |
| 7 | 6 | 4 | 7 (92) |

**Table 6**

**An Instance of a Maximum Size (6; 3)-Error Correcting Code Among those Locate by the Evolutionary Algorithm. This Code has 106 Members, All at Mutual Edit Distance at Least 3. The Unmodified Version of Conway's Lexicode Algorithm Locates a 96 Member Code.**

| | | | |
|---|---|---|---|
| GTGCTC | ATTGGC | ACGGOG | CGOCTG |
| GACTAA | AGGAGC | GAAGOG | ATACTG |
| OOCAGC | TAGTGC | TTGACG | GTTGTG |
| GOCOOC | ACATGC | GCTAOG | COGATG |
| CGGOCC | ATCCAC | GTCOGG | TGAATG |
| AAAOOC | TAOGAC | CAGCGG | AOCTTG |
| TTTCOC | CTGGAC | AGAGGG | TATTTG |
| AGOGCC | GGCAAC | TCTGGG | TGACCA |
| TCAGOC | AATAAC | AACAGG | CTGGCA |
| CAGACC | CGATAG | CGTAGG | CGCACA |
| CTAACC | TCTTAC | GGGTGG | ACAACA |
| CACTOC | TGGGTC | TTATGG | TATACA |
| ATGTOC | CCTGTC | GOGCAG | TOCTCA |
| GGTTCC | TTCATC | AGTCAG | CTTTCA |
| TGOCGC | GAATTC | CATGAG | ATGOGA |
| GATOGC | TAOCOG | TOCAAG | ACOGGA |
| GOGGGC | CCAOOG | GTAAAG | CGGGGA |
| CAAGGC | CTOGCG | AAATAG | TCGAGA |
| ATTGGC | ACGGCG | CGOCTG | GGAAGA |
| AGGAGC | GAAGOG | ATACTG | GCTTGA |
| TAGTGC | TTGACG | GTTGTG | OOCCAA |
| ACATGC | GCTAOC | COGATG | TAGCAA |
| ATOCAC | GTCCGG | TGAATG | GTTCAA |
| TAOGAC | CAGOGG | AOCTTG | GCAGAA |
| CTGGAC | AGAGGG | TATTTG | TGTGAA |
| GGCAAC | TCTGGG | TGACCA | CAAAAA |
| AGGAGC | GAAGOG | | |

The application for the error correcting codes in the edit metric is to provide embeddable bar codes for cDNA libraries. Because these bar codes are to be embedded in constructs there are a number of constraints on the sequence that may be used that are driven by biology beyond the need for error correction. In making the constructs various restriction enzymes are used which cut a DNA strand at a particular pattern. We must avoid creating additional instances of this pattern either within our bar codes or as a side effect of embedding our bar codes into the construct. Sub-strings of the form TT or AAA will interfere with use of the construct because of a long sting of T's near the point were the bar code is embedded.

It turns out that modifying the evolutionary algorithm to deal with such constraints is not difficult. In the seed generator, mutation operators, and in the greedy algorithm, a short piece of code is called that checks the acceptability of each string relative to the biological constraints.

## 6. SUMMARY

Three examples of applications of evolutionary computation to bioinformatics are presented here. The first differs from the simple introductory example, the string evolver, only by having a different fitness function. With a fitness function that minimized the randomness of an alignment of a set of 315 sequences the string evolver was able to bring a motif correlated with transposon insertion over the threshold of detectability. While quite simple as an example of evolutionary computation, this application solved a real problem. Because of the essential simplicity of evolutionary computation, the entire software development effort needed to solve that problem took an afternoon.

The creation of finite state machines to learn how to second-guess a standard primer-picking package was a more difficult effort. A large amount of data, in the form of scored primers, was required before the effort began. The choice of a representation was not as obvious as in the first example. Finite state machines are well able to pick out a pattern no matter where it appears along the length of the primer and so were selected as the representation for our classifiers. The standard fitness function—scoring the number of correct predictions—did not work well in preliminary studies. The use of an incremental reward, computed as a primer traverses the finite state machine, worked well enough to have an impact on future costs. Classifiers located with the incremental fitness function will decrease the number of bad primers used. Substantial room remains for improvement. The evolutionary algorithm used to evolve finite state machines to classify primers (equivalently: to learn the patterns in the scored primer training set) was a fairly standard evolutionary algorithm. The only feature not completely standard was the fitness function. It is also worth reminding the reader that this evolutionary algorithm over-trained the finite state machines when permitted to hybridize a second time. It is not possible to over-emphasize the need for cross validation when learning from data.

The creation of larger sets of error correcting DNA bar codes used a new type of evolutionary algorithm called a greedy closure evolutionary algorithm. The basic notion is to first choose a greedy algorithm that extends partial structures. In this case Conway's lexicode algorithm is the greedy algorithm. The representation for this type of evolutionary algorithm is a small initial part of the structure, in this case a seed of three initial DNA bar codes. The fitness is the quality (in the case of error correcting codes: size) of the final structure constructed by the greedy algorithm. While we used this technique to find larger sets of DNA bar codes it has many other possible applications.

These three examples, while quite different from one another, do not do justice to the breadth of evolutionary computation. Evolutionary computation has been used since at least the 1960s (Fogel, Owens, and Walsh, 1965) with techniques similar to those used in the primer-picking example. Foundational works in the area include (Goldberg, 1989; Holland, 1992) which introduce a type of evolutionary computation called *genetic algorithms*. Evolution of variable sized structures, including whole computer programs, comes under the name of *genetic programming*

(Kinnear, 1994; Kinnear, and Angeline, 1994; Koza., 1992 and 1994). While these techniques have in common the basic structure given in Figure 1, they each incorporate unique features and potential pitfalls.

Evolutionary computation started as a machine learning and optimization techniques, having been discovered many times in many places. It was not practical until the late 1980s when the size of widely available computers grew to where it could support the long run times required. The algorithms used by evolutionary computation are fast to write, slow to run, and easy to specialize for particular tasks. In general, pure evolutionary computation does not perform well on problems that have been studied for a long time. This is because pure evolutionary computation is too simple to take advantage of expert knowledge about problems. Hybrid evolutionary algorithms, where evolutionary computation is blended with other techniques, can incorporate expert knowledge and does often compare well with or beat other techniques. The DNA bar codes are an example of a high performance hybrid technique.

Evolutionary computation is an option for problem solving best used in the initial, exploratory stages of a project. Algorithms that are less general-purpose can almost always out-perform evolutionary computation. Such specialized algorithms, while they supplant evolutionary computation techniques, may require knowledge gained with initial studies of the problem that used evolutionary computation. In summary, evolutionary computation is so easy to use that it is a good choice for brain-storming and prototyping. It is also quite a lot of fun.

## 7. REFERENCES

Ashlock, D.A., and Joenks, M., 1998, ISAc lists, a different representation for program induction. In *Genetic Programming 98, Proceedings of the Third Annual Genetic Programming Conference*, pages 3–10. Morgan Kaufmann.

Ashlock, D.A., Wittrock, A., and Wen, T.-J., 2002, Training finite state classifiers to improve pcr primer design. In *Proceedings of the 2002 Congress on Evolutionary Computation*, pages 13–18. IEEE Press.

Ashlock, D.A., 2004, *Optimization and Modeling with Evolutionary Computation.* Springer-Verlag, Inc. New York, NY.

Baldi, P., and Brunak, S., 2001, Bioifnormatics, the Machine Learning Approach, second edition. MIT Press, Cambridge, MA.

Dietrich, C., Cui, F., Packila, M., Ashlock, D., Nikolau, B., and Schnable, P.S., 2002, Maize mu-transposons are targeted to the $5^1$utr if the gl8a gene and sequences flanking mu target site duplications throughout the genome exhibit non-random nucleotide composition, Genetics, **160**:697–716.

Fogel, G.B., and Corne, D.W., 2003, *Evolutionary Computation in Bioinformatics.* Morgan Kaufmann Publishers, Boston, MA.

Fogel, L.J., Owens, A.J., and Walsh, M.J., 1965, Artificial intelligence through simulated evolution. In *Biophysics and Cybernaetic Systems: Proceedings of the 2$^{nd}$ Cybernetic Sciences Symposium*, pages 131–155.

Goldberg, D.E., 1989, *Genetic Algorithms in Search, Optimization, and Machine Learning.* Addison-Wesley Publishing Company, Inc., Reading, MA.

Holland, J.H., 1992, *Adaption in Natural and Artificial Systems.* The MIT Press, Cambridge, MA.

Kinnear, K., 1994, Advances in Genetic Programming. The MIT Press, Cambridge, MA.

Kinnear, K., and Angeline, P., 1994, *Advances in Genetic Programming, Volume 2*. The MIT Press, Cambridge, MA.

Koza., J.R., 1992, *Genetic Programming*. The MIT Press, Cambridge, MA.

Koza,. J.R., 1994, *Genetic Programming II*. The MIT Press, Cambridge, MA.

Peter, J., 1997, Angeline. Subtree crossover: Building block engine or macromutation? in: *Genetic Programming 1997: Proceedings of the Second Annual Conference*, J.R. Koza, K. Deb, M. Darigo, D.B. Fogel, M. Garzon, H. Iba, and R.L. Riolo, eds, pages 9–17. Morgan Kaufmann, 1997.

## Chapter 3

# Architectures for Integration of Data and Applications

## Lessons from Integration Projects

William D. Beavis

## 1. INTRODUCTION

Despite the hyperbole and excitement about bioinformatics in the late 90s, during the last two years bioinformatics as a stand-alone commercial enterprise has failed. Of dozens of Bioinformatics companies in existence in 2000, only a few remain in early 2003 (Toner, 2002). This is due in part to the overall economic malaise and the "bust" of technology-based companies. However, there are fundamental reasons that are more complex. Consider, for example chem-informatics companies have not only survived during this same time, but have actually thrived (Cramer, R. V.P.for Research of Tripos Inc., personal communication). The difference is that chem.-informatics activities are very focused on discovery of small molecules that will have biological activity. Specifically, chem-informatics is focused on development of databases and algorithms that will search through libraries of potential (and existing) small molecules that could bind DNA and proteins and thus serve as drug candidates. Chem-informatics is recognized by investors as focused on discoveries of potential products with short-term return on investment. In contrast, Bioinformatics encompasses a broad set of activities involved in a wide range of discoveries from candidate genes responsible for complex phenotypes

**William D. Beavis**   National Center for Genome Resources, Santa Fe, New Mexico, 87505

*Genome Exploitation: Data Mining the Genome*, edited by J. Perry Gustafson, Randy Shoemaker, and John W. Snape.
Springer Science + Business Media, New York, 2005.

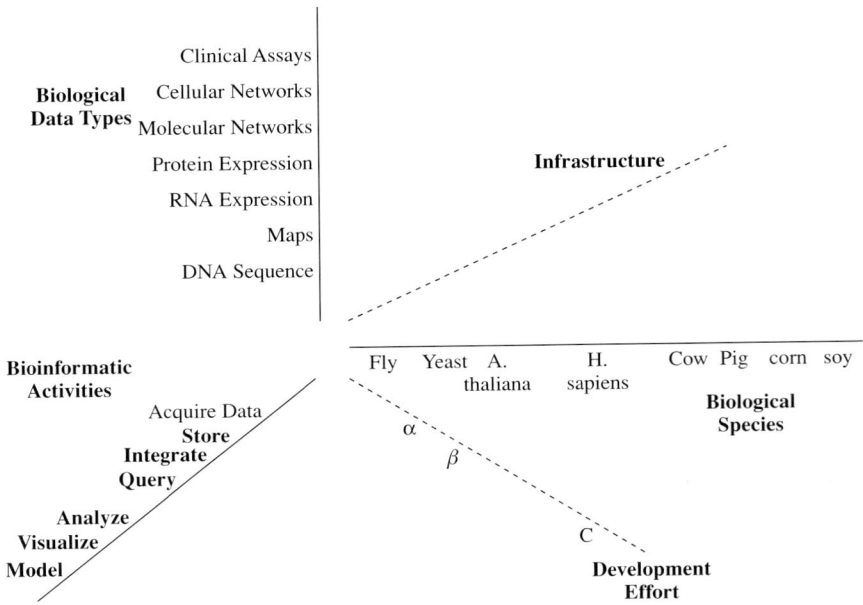

**FIGURE 1.** Representation of the research and development space spanned by various aspects of Bioinformatics.

to understanding the structure and evolution of whole genomes. Even the more focused and applied bioinformatics goals, e.g., discovery and characterization of binding sites for expression of candidate genes and proteins, are recognized as long-term investments.

There are, however, positive outcomes to the more fundamental and broadly defined goals of Bioinformatics. First, Bioinformatics is emerging as a scientific discipline. There are now over 150 Bioinformatics programs at US universities. Second, although Bioinformatics may not be viable as a stand-alone commercial enterprise, it has emerged as an integral part of successful commercial life-science companies. Virtually all now have large Bioinformatics Departments because they recognize their long-term viability will depend on fundamental discoveries that map the relationships between genotypes and phenotypes.

At a high level, the Research and Development Space of Bioinformatics can be viewed as a set of non-orthogonal vectors (Figure 1) that describe

> *Bioinformatic Activities*
> *Biological Data Types*
> *Biological Species*
> *Computing Infrastructure*
> *Development Effort*

*Bioinformatic activities* (acquisition, storage, retrieval, integration, analysis, visualization, modeling) need to be developed for multiple *biological data types* (nucleic and amino acid sequences, physical and linkage maps, RNA, protein and metabolite expression arrays and clinical and field assays) derived from multiple *biological species* using multiple biotechnology platforms. Notice that the scope of bioinformatics exceeds existing resources and capabilities of most organizations including large pharmaceutical companies. Thus, decisions about the amount of *development effort* (theory, prototype, alpha, beta, production) and *computing infrastructure* that a Bioinformatics program is going to devote toward the development of these tools and methods need to be made based on the planned strategic directions of funding agencies, as well as the expectations, interests and competencies of the relevant faculty or staff.

With the emergence of numerous high throughput biotechnologies that produce large amounts of various "-omics" data, there is widespread recognition for the need to develop integrated software systems. The problem space of integrated software systems for bioinformatics can be categorized according to content and approach. That is, in order to address an integration problem in bioinformatics, one needs to understand what data, information and applications need to be integrated as well as the possible technical approaches that can be pursued; recognizing that the development of an integrated software system is likely to transcend multiple categories in both of these dimensions.

The simplest form of integration is to combine data from repeated or related experiments for either combined analyses or meta-analyses. Once able to combine data the next logical thing to do is integrate data with analysis tools and to integrate multiple analysis tools so that the output from one analysis can serve as input to subsequent analyses. More complex forms of integration across species, data-types and applications are necessary for the synthesis of knowledge (Seipel et al., 2001a). For example, integration of information derived from unrelated experiments across data types provides the researcher with the ability to validate and narrow the search for candidate genes. Consider also the field of comparative genomics where data and information from multiple species are integrated for purposes of understanding evolutionary relationships.

Informatics has provided us with a large number of technical approaches to integration. Seipel et al. (2001a, 2001b) described these approaches as either top-down or bottom-up (Figure 2). Top-down approaches can be characterized by their coordinated development of the database, interfaces, and applications, use of well defined standards for data representation and data transfer protocols, and homogeneous data. An example is the data warehouse where all information on a species is served through a database and interfaces that are developed as a single system. The advantages include tight, consistent and high quality integration of all data types (Ritter, 1994), thus allowing very efficient, complex queries and effective representations of multi-dimensional data. The disadvantages include high costs for development and inability to incorporate emerging data types and applications because every change to any single aspect of the system will have, often unanticipated, impacts throughout the system.

Thus, in rapidly evolving scientific disciplines, such as bioinformatics, one is faced with abandoning systems before they are completed. Bottom up approaches

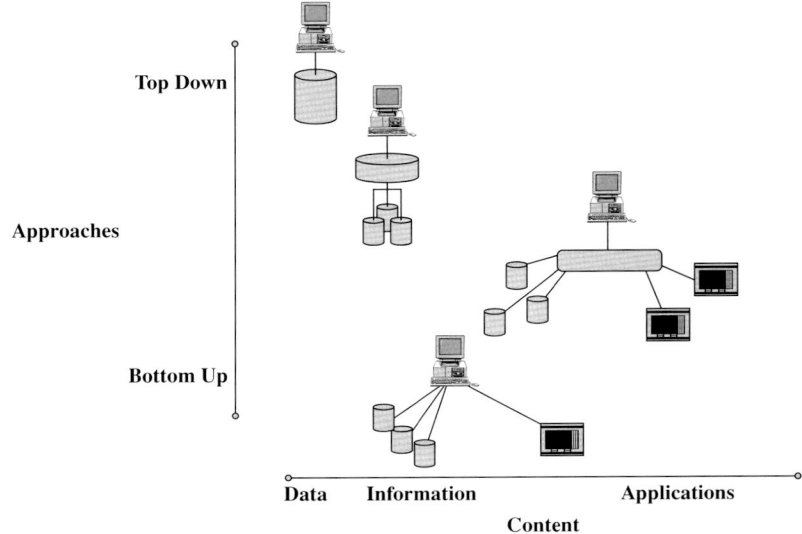

**FIGURE 2.** Representation of approaches to integrate heterogeneous content, i.e., various aspects data, information and application (query tools, analyses, visualization tools).

are characterized by independent development of databases and applications as components with little adherence to a single set of standards for data representation and data transfer protocols. Rather than attempting to homogenize the data, heterogeneity is embraced (Benton, 2000). The use of web-enabled hyperlinks represents a well-known example of this concept, although it could be argued that hyperlinks do not represent actual integration. There are also component-based approaches that fall between these extremes. Application service providers, federated databases, and component-based integration represent examples of these intermediate approaches. Component based approaches do not produce systems capable of complex declarative queries because they do not provide inter-database consistency. However, they do allow heterogeneous resources to exchange data without homogenizing the data and applications (Szyperski, 1998). Thus providing development and maintenance cost advantages (Searls, 1995; Goodman, et al., 1995; Boyle, 1998; Szyperski, 1998) as well as greater flexibility in development of intuitive user-interfaces (Searls, 1995; Boyle, 1998; Fischer, et al., 1999). While each of the component-based approaches have their own advantages and disadvantages (Siepel et al., 2001b), a common feature includes the ability to adopt new and changing data types and applications from diverse sources. This is particularly useful in rapidly changing scientific disciplines where novel technologies and methods cannot be predicted (Gessler, 2002).

When I first arrived at NCGR over four years ago there was a great deal of interest, scholarship and debate among our biologists and software developers about the various approaches to integration. We've had a chance to investigate and apply theory through development of numerous integrated systems including the Genome Sequence Database (GSDB), the Arabidopsis Information Resource (TAIR), X-Genome Initiative (XGI), GeneX, PathDB, ISYS, MOBY, and the Legume Information System (LIS). To varying degrees of success, these have been implemented across biological data types and biological species in several computing environments. Some of these projects have been quite successful in terms of technical innovations, but unsuccessful in terms of wide-spread use by biologists, while other projects have been rather mundane in terms of computer science and technical innovation, but very successful in terms of use by biologists. In the following, I will describe four of these integrated software systems (TAIR, XGI, GeneX and ISYS) and the lessons we learned from these development projects. The single most important lesson is not about the relative technical merits of the approaches rather it is about the essential participation of biologists as partners and collaborators.

## 2. A WEB-ACCESSIBLE DATA WAREHOUSE (TAIR)

### 2.1. Description

The development of The Arabidopsis Information Resource (TAIR) is an example of a top down approach to integrating all known data, information and applications for a single model plant species, *Arabidopsis thaliana*. It is a web-accessible data warehouse that we have been developing in collaboration with plant biologists at The Carnegie Institute of Washington at Stanford University (Huala et al., 2001). The primary goal of TAIR is to provide web access to all research information about Arabidopsis and thus promote interactions in the plant research community. As a database system TAIR integrates all of the known genomic, genetic, molecular, biochemical, metabolic, pathway and phenotypic information for a single species, although the database schema is sufficiently robust to accommodate multiple species. The data held in TAIR is acquired in a number of different ways. Data provided by genomic sequencing centers like TIGR deliver data in bulk format, which are parsed and loaded directly into the database. Other data are acquired automatically from online database sources like GenBank. Other data are provided directly by researchers and this data is processed by either ABRC or the TAIR curators and sent as files to the database for loading. Finally, a significant amount of data is extracted from publications by the curators and sent via a pipeline to the database for automated loading. In the near future TAIR will also allow individuals to dynamically upload data for storage via the web site. Perhaps TAIR's greatest asset is the careful curation of data by plant biologists under the supervision of Dr. Sue Rhee at the Carnegie Institute.

The current version of TAIR actually represents a third generation in the evolution of the system. The original system, known as AtDB, was developed by Mike Cherry at Stanford University (Flanders et al., 1998). The second generation was conceived in a proposal (Huala et al., 2001) to NSF that currently supports development of TAIR, but was short-lived because we learned that our original concepts for the architecture were not sufficiently robust to support the amount and diversity of data or the number of queries from the research community. Since autumn of 1999, when NSF began funding TAIR, the size of the database has grown to support about 29,000 genes, 400,000 nucleotide sequence entries associated to data objects, 200,000 mutant lines, 4,000 genetic markers mapped on the sequenced genome, 90,000 polymorphisms mapped on the genome or associated with germplasm resources, 500 microarray experiments, 14,000 publications, 11,000 researcher and 4,000 organization profiles (ref NAR 2003 and http://arabidopsis.org/about/). Furthermore, there has been an increase from 20,000 web page visits per month in November 1999 to over 500,000 per month (http://arabidopsis.org/usage/).

From a computer science perspective, TAIR is a software development project with very little research. In order to accommodate a maximum diversity of platforms and operating systems used by plant biologists TAIR is served through a trio of servers: a web server that processes all incoming requests; a database server that houses all the Arabidopsis data; and an analysis server which handles computationally intense data processing tools like BLAST and Patmatch. TAIR is accessible at (http://arabidopsis.org) through commonly used web browsers. The software is written in Java, Javascipt and HTML with an emphasis on minimizing platform dependency issues experienced by the various web browsers and the hardware systems they are operated on, and is accessible at (http://arabidopsis.org) through commonly used web browsers. The design of the database and application tools is based on an object-oriented approach. The database is implemented using Sybase (version 12.5), a relational DBMS. We designed flexibility into the database by minimizing linkage among tables. The data tables are organized in a hierarchical structure where attribution, reference and annotation classes constitute meta-data of all TAIR objects (http://www.arabidopsis.org/search/schemas.html). As a result we have ensured optimal responses to queries as the system grows to accommodate new data types from emerging technologies.

All software applications are implemented in a client-server mode also using the JAVA Servlet technology and are accessible to researchers by common web browsers. Information can be retrieved and visualized in a number of different ways. For example, SeqViewer and its complement, MapViewer, provide views of the genome decorated with genes, clones, transcripts, genetic markers and polymorphisms in the context of whole chromosomes and chromosomal segments and can be used to compare Arabidopsis sequence, physical, and genomic maps. Alternatively, genes can be viewed in the context of metabolic pathways and biochemical structures using AraCyc.

TAIR also provides some analysis services including BLAST, FASTA and Patmatch, which provide sequence similarity analyses against Arabidopsis and plant-specific sequences. Matching sequences are hyperlinked to the TAIR loci as well as to the MIPS and TIGR databases and can be used to retrieve other associated data such as GO annotations. However, because researchers often want to conduct their own analyses a set of bulk data download tools that produce flat files also are available (http:// arabidopsis.org/tools/bulk/).

## 2.2. Lessons

Despite use of well-established design principles and a state of the art DBMS to assure simplicity and flexibility, TAIR is now a very complex and large system. Changes to the system to accommodate new data types or applications are more and more difficult to implement. This, of course is a well-recognized consequence of utilizing the single data warehouse concept (Siepel et. al., 2001a), but was not fully understood at the time we proposed the project. From a financial perspective TAIR is not our most successful project; we have subsidized as much as $1/4$ of the direct labor costs for the project. Also, because TAIR is a development project, there have not been many peer-reviewed manuscripts generated in the course of the project. None-the-less, with respect to use and impact on the biological community TAIR is the most successful integrated software system that we have developed. Growth in both content and use are measurements of this success.

Although development of such a large complex system depends on technical expertise, our experience with this system and others (described below) is that success depends more on clear and good communication between biologists and software developers. From a technical perspective it is the clearly defined, structured and shared vocabularies between software developers and biologists that were essential to design of the system. It has been the responsibility of Dr. Sue Rhee to assure that this communication takes place despite the physical distance between Santa Fe and Palo Alto as well as the cultural distance between biologists and software developers.

In many respects development of this resource is as much about technical considerations as it is about building a sense of community. This not only requires good communication between software developers and biologists on the project team, but also active communication with the broader Arabidopsis and other model organism research communities. To assure this broader participation, Sue has represented TAIR as an active member of the Gene Ontology Consortium (http://www.geneontology.org) since 2000 and has provided our perspective on gene models and the roles of genes and products so that it will be possible for biologists from different backgrounds to investigate similarities among diverse organisms.

## 3. AN AUTOMATED ANALYSIS AND ANNOTATION PIPELINE (XGI)

### 3.1. Description

Like many organizations that have been faced with large continuous streams of DNA sequences from multiple projects, we have automated analyses and annotation. In an effort to consolidate these efforts into single automated process we developed XGI (Genome Initiative for species X). The goal of the XGI project has been to develop a system that is powerful enough to process large amounts of both genomic and expressed sequence data and flexible enough to incorporate different analysis tools depending upon the needs of a particular project. The

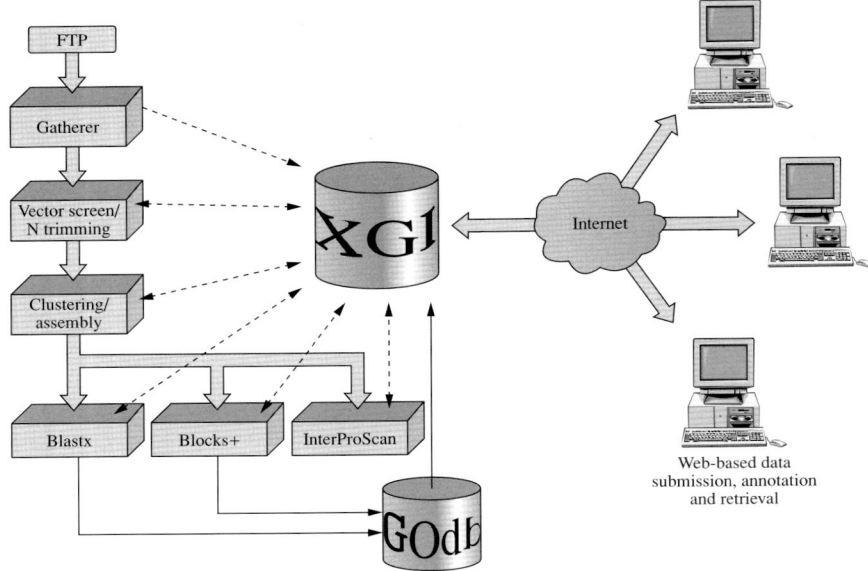

**FIGURE 3.** Representation of the architecture for automated analysis, annotation, storage and retrieval of sequence information used by XGI.

components of XGI can be described as three interacting modules: an analysis pipeline, a relational database and user interface (Figure 3).

In the context of research and development, XGI, like TAIR, is a software development project, although the results of the analyses are used extensively for comparative genomics research. Development of the pipeline might be viewed as utilizing a bottom up approach because most of the analysis and annotation tools as components that were developed independently of the pipeline. In contrast, the database, user interface, and data transfer protocols were all developed using a top down approach. The analysis pipeline is fully parallel and can be run in symmetric multiprocessing mode (SMP) on shared memory machines, or can be run in distributed multiprocessing mode (DMP) on clusters of workstations or on Beowulf clusters. It uploads raw sequence data and processes it through a series of analysis operations, each operation building upon the results of the previous stage. Results of each analysis stage in the pipeline are stored in the XGI relational database that has been implemented in both Sybase (v 12.5) and Oracle (v 9i ). The web interface provides the ability to pursue powerful complex declarative queries using Boolean operators on the part of our collaborators. The XGI Web Interface provides researchers with the ability to submit complex queries that permit precise delineation with Boolean logic operators in conjunction with scope delimiters (STRICT, LOOSE) enabling virtual Northerns and *in silico* subtractions as well mine the results of the analyses based on keyword searches of features and GO annotations.

The analysis pipeline is fully parallel and can be run in symmetric multiprocessing mode (SMP) on shared memory machines, or can be run in distributed multiprocessing mode (DMP) on clusters of workstations or on Beowulf clusters. It uploads raw sequence data and processes it through a series of analysis operations, each operation building upon the results of the previous stage. Due to its modular design, it is possible to add sequence analysis tools to the pipeline through several mechanisms including: writing the new analysis in JAVA, use of the JAVA Native Interface (http://java.sun.com/docs/books/tu-rial/native1) for tools written in C or C++, and use of XML (http://www.w3.org/XML/) to incorporate programs written in other languages. Results of each analysis stage in the pipeline are stored in the XGI relational database that has been implemented in both Sybase (v 12.5) and Oracle (v 9i). The XGI Web Interface provides researchers with the ability to submit complex queries that permit precise delineation with Boolean logic operators in conjunction with scope delimiters (STRICT, LOOSE), thus enabling virtual Northerns and *in silico* subtractions as well mine the results of the analyses based on keyword searches of features and GO annotations.

The current version of XGI has been through several generations of development that have been supported by the Novartis Foundation, the Noble Foundation, USDA-ARS, NSF-IFAFS, NSF-PGRP and UC-Davis. These organizations represent several diverse biological research communities with interests in the plant pathogens *Phytophthora infestans, Phytophthora sojae*, several species of legumes including the reference legume *Medicago truncatula*, and cotton. Since its inception the pipeline has processed hundreds of thousands of ESTs from several plant and fungal species and thousands of genomic BACs and Contigs, primarily from *Medicago truncatula*. When utilizing all steps in the pipeline it is currently capable of processing about 2000 ESTs per day or 300 BACs per day on our 16-node cluster. By far the slowest stage in the process are the seven searching steps performed by InterproScan, which constitute about 40% of the time for an EST pipeline run. This is due to several computationally intensive hidden markov chain-based algorithms. It is possible to run the pipeline without sending the sequences through the InterProScan stage (or any other stage for that matter) by using an Administration Tool, which also makes it possible to adjust the parameters associated with any of the analysis steps. Because XGI achieves high-throughput by parallelization, performance will improve in a near linear fashion as the number of available processors increase.

## 3.2. Lessons

XGI has been a successful integration project. From a financial perspective, it has been fully supported by the communities that it serves. On the other hand, it has not produced many peer-publications and it has not been widely used by biological researchers outside the collaborative research communities for which it was designed. Results of analyses are accessed primarily by the collaborative research communities involved in the specific genome sequencing projects.

Although developed for use by specific collaborative genomics research projects, the XGI system is portable and can be installed offsite. It is available through a no cost license to academic researchers and it has been successfully installed and is in use at Plant Research International in The Netherlands. None-the-less, to date there have been few external installations of the software system. I think that there are several reasons for this. First, until recently we did not devote resources into development of a licensing model. Second, XGI was designed and developed by bioinformaticists to meet the need for automated analyses and annotation of specific collaborative projects, not as commercial software for distribution. Thus, installation of XGI requires computational expertise to set up and run. Third, it was developed to process large amounts of sequence data from collaborative genomics projects and to present the results to remote sites through web interfaces. It was not designed to be a stand-alone desktop application. For this reason it is not suited to running multiple iterations with small datasets, which has been an oft-asked request from biologists. Finally, most large-scale sequencing projects have developed their own pipelines. Likewise smaller projects have built smaller scale pipelines to meet their specific needs with students and post-docs.

## 4. AN OPEN SOURCE INTEGRATED SYSTEM (GENEX-LITE)

### 4.1. Introduction

Several years ago we initiated a large project, which we named GeneX, to integrate data and information from gene expression arrays. Our original intent for GeneX was to develop a system that would integrate data and information across technology platforms and across biological species with the flexibility to be deployed at an individual lab, groups of labs and globally (Figure 4). Similar to TAIR, our approach was top down with the goal of developing and supporting a global warehouse to support comparative functional genomics. In contrast to

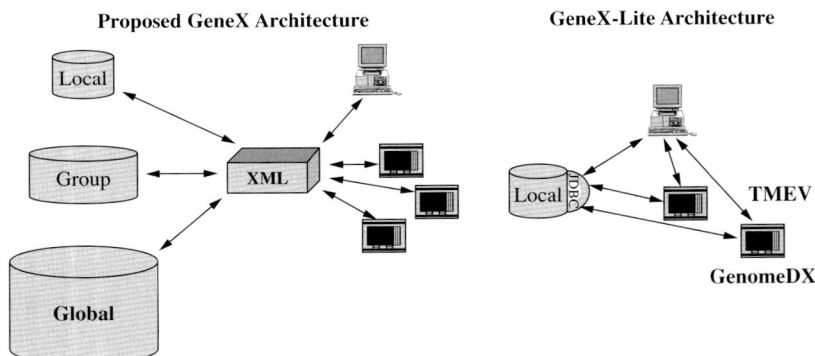

**FIGURE 4.** Comparative representation of architectures used by GeneX and GeneX-Lite.

TAIR, we were interested in a single data type rather than a single species. Also, we wanted to manage the development through an open source model and take advantage of emerging standards for modeling and transfer protocols of data from gene expression experiments.

We planned the system to support data generated by micro-arrays (Schena et al., 1995), Affy-Chips (Lipschutz et al., 1999), nylon arrays, AFLPs (Vos et al., 1995) and Serial Analyses of Gene Expression (Velculescu et al., 1995) with

- an open-source relational DBMS,
- an extensible DB schema
- scripts and applications to handle data input and output
- code to interface the DB with applications
- statistical analysis applications
- and a client-side Curation tool with which we expected to curate
- a "large number" of publicly available expression arrays.

We initiated this project with internal resources, but also obtained funding from NSF after about a year of effort. The reviewers at NSF were fairly skeptical of the scope of the original proposal and strongly encouraged us to reduce the scope to show proof of concept; it was advice that we did not fully appreciate at the time.

In the first two years of the project we learned a great deal about how *not* to manage a large software development project, particularly in a dynamic environment of constantly changing data and information. At the end of our second year (first year of NSF support) we had developed relatively large and complex relational data model that was capable of accommodating virtually all types of expression data along with most associated data describing species, experimental design, gene annotation, etc. The database was implemented in Postgres, an open source RDBMS. We felt very good about this accomplishment given the evolving standards during the time we were developing the system. However, with the release of MIAME, we realized that the schema was unnecessarily complex. We also developed an XML based data transfer protocol and implemented it in an object layer between the database and applications. The data model for this middle layer was also quite complex to accommodate the anticipated diversity of data. Unfortunately, the data model for this middle layer did not map well to the data model of the database. Finally the system had several applications consisting of a curation tool, an upload tool, a web interface, and several analysis and visualization applications including R, CyberT (xgobi and xviz), J-Express and xcluster. The weakest applications at the time were the curation and upload tools, which required researchers to fill out several pages of information in order to upload even simple data sets. Indeed our biologists only managed to upload a couple of publicly available datasets in a six-month period.

The entire GeneX system was released initially through SourceForge (http://genex.sourceforge.net) and subsequently through the NCGR web-site (http://genex/ncgr.org/download). Although functional, the entire system was cumbersome. Contrary to our expectations, releasing the code through an open source model did not attract software developers interested in joining the development process. I think there are a couple of reasons for this. First, as Peter Rojas (2000) points out, simply putting code on a server with an open source license is not sufficient for success. The code needs to be cleaned up and well documented before the open source community is willing to work with it. Second, the code must address a need recognized broadly by software developers. There are not large numbers software developers with sufficient knowledge or interest in gene expression; those with sufficient domain knowledge are probably already working on their own systems.

The code was downloaded primarily by biologists, but they found the system particularly the curation and upload tools, to be too cumbersome to meet their needs. Most downloads by biologists were motivated by a desire for a free software system that could be utilized for management and analysis of expression data in the individual laboratory setting. We therefore decided to develop a simpler system, based on the MIAME standards (http://www.mged.org), to meet the needs of the single investigator. We refer to the system as GeneX-Lite due to the simple, somewhat abstract, data model and simpler interface design (Figure 4). The new schema easily supports a wide variety of experimental designs and data from multiple technology platforms (http://www.ncgr.org/genex/architecture.html). Thus, it is possible to combine selected data from different experiments into an *in silico* or virtual experiment; assuming that the data are properly normalized before analyses.

The entire system including database server, web server, and all applications can be run on a laptop and has been installed in UNIX, windows and OSX environments. We also abandoned the XML mechanism for data transfer, replaced it with JDBC and developed a data upload mechanism that is flexible and fast. It is flexible in the sense that it is possible to load and annotate information and data in separately and incrementally. Individual research biologists have found this to be quite useful because annotations and meta-information can be added as needed. Our own bioinformaticists have loaded over 520 experiments into the system; the largest of these by Kim et al. (2001) had almost seven million measurements and took only about 15 minutes to load.

Most recently, with the open source release of TMEV (http://www.tigr.org), we decided to see if the code from both open source systems could be integrated. It turned out to be very easy and straight-forward. We then discarded our efforts to integrate CyberT and the other analysis and visualization tools because TMEV provided greater functionality with a single interface. We also abandoned the xgobi/xgvis visualization tools in favor of those provided by TMEV because xgobi/xgvis rely on X-Windows and can be troublesome on some platforms with network configurations. In the near future we will be extending the visualization capabilities of the system by incorporating GenomeDX (Matthews et al., 2002), an open source visualization and data-mining tool.

Since completing integration between GeneX-Lite and TMEV the number of down loads of the system has increased dramatically. Currently, GeneX-Lite registrations average about 60 per month. These are primarily by biologists and bioinformaticists. None have contributed code, but many have made suggestions to improve the user interface, which have been easy to implement.

## 4.2. Lessons

GeneX has been our most difficult software development project, while GeneX-Lite has become one of our more successful projects. Our original concept of deploying a comprehensive system at individual labs, for a group of labs or globally through an open source model was naïve. We missed all three targets with our first few releases of the GeneX system and failed to attract open source developers. Since changing our development model to focus on the needs of a single investigator while paying close attention to emerging standards for data modeling and exchange, we have developed a functional system that seems to be meeting the needs of many.

There are numerous reasons for lack of success with the initial GeneX project. Recall that top down approaches are characterized by their use of well-defined standards for data representation and data transfer protocols through coordinated development of the database, interfaces, and applications. While we had members of our development team participate in development of MAML and subsequently MIAME standards these standards were not settled during our first two years of development. Subsequent development of GeneX-Lite has benefited from the release of MIAME and will benefit from various MAGE initiatives dealing with data exchange protocols.

More importantly, the development of GeneX modules was not well coordinated. Because we were enamored with the open source philosophy, but did not understand the open source development model, our developers tended to interpret the open source model as permission to pursue a bottom up approach to development. Thus, the modules tended to be developed independently. With turnover of project staff, however, we had an opportunity to reevaluate our approach and realized that top down approaches were not incompatible with an open source model. Indeed, they are quite compatible, as evidenced by the coordinated development of LINUX through strong management by a "benevolent dictator".

Most importantly, rather than try to develop a comprehensive system for our perception of all researchers needs, we decided to focus on development of an integrated system that could be deployed into the lab of single researchers and listened to a single yeast researcher, Dr. Stephanie Ruby, at the University of New Mexico Health Sciences Center.

## 5. LOOSE COUPLING INTEGRATION (ISYS)

### 5.1. Description

Several years ago we began a component based approach to the development of an Integrated System (ISYS$^{TM}$) that would provide greater integration than simple hyperlinks yet still facilitate an exploratory environment for the researcher. We were motivated by several factors. First it was our sense that Biologists would prefer to explore relationships in databases rather than specify complex queries *a priori*, i.e., most researchers would prefer to compose queries interactively while probing, inspecting and exploring database information. Thus, we wanted to build a system that facilitated biological discovery. Second, we recognized that bioinformatics is a rapidly changing discipline where novel methods and technologies are continually emerging. Thus, we wanted to build a flexible integration platform that would allow researchers to adopt new and changing data types and applications from diverse sources. Although funded primarily with internal resources, we also obtained funding and collaborative support from four members of the Consultative Group of International Agricultural Research (CGIAR) centers who felt that our implementation as a client-side integration system could be beneficial to their programs consisting of widely distributed researchers with poor and inconsistent connections to the internet and other centralized information resources.

Our approach with ISYS was similar to that advocated by the Life Sciences Task Force (1997) of the Object Management Group in which heterogeneous components, including both databases and applications, can interoperate, but allow freedom in the assembly of systems from components (Slidel, 1998; Benton, 2000). The resulting integration platform is capable of integrating websites, databases, service providers and applications that have been developed independently of ISYS. Integrating components into the platform is straight-forward using standard protocols and is published in the ISYS API (http://www.ncgr.org/isys/developers). To date, we have integrated over a dozen components, including sequence viewers, the Berkeley Drosophila's Genome Project's gene ontology browser, maxdView's gene expression viewer, interfaces to local and remote BLAST, and numerous web pages. The integration of web pages allows researchers to invoke web pages as services and extract results from web pages to bring them back client-side for further analysis.

Details of how to implement ISYS are published in the ISYS SDK (Software Development Kit) available at www.ncgr.org/isys/developers.html/. Features of the system include synchronous display among components and the ability to discovery appropriate services based solely on the type of data they accept, which we refer to as DynamicDiscovery. These are direct outcomes of the use of an Event Channel and Broker Pattern in the ISYS architecture. The event channel allows components to "listen and react" to other components thus providing synchronous displays of filtered information. Our implementation of the Broker Pattern in the ISYS bus allows all components to request and provide services to and from all other components, thus allowing independently developed components to dynamically volunteer their services on selected objects.

ISYS has been available to the research community for a couple of years and over 3200 copies have been distributed. It is not clear to us whether other researchers are adding their own choices of components to the system. We have been working in partnership with molecular plant breeders at the Consultative

Group of International Agricultural Research (CGIAR) centers to develop and add components of their choosing to the system.

## 5.2. Lessons

From a scientific and technical perspective we managed to bring two powerful features, dynamic discovery and component synchronization, to the problem of integrating heterogeneous and independently developed bioinformatics resources. From a financial perspective, most of the direct labor costs of the research components of the project have been funded with internal resources. And from a "use" perspective, we are not sure how many biological researchers are using the system. From our collaborative efforts with the CGIAR researchers it seems that synchronization and dynamic discovery are features that researchers have been very useful for probing and exploring multi-dimensional information resulting in discovery and synthesis of knowledge. At the same time, it needs to be noted that there are several learning curves that need to be engaged when using such a system. First, the set of components that we have integrated may not be familiar to the biologist. Second, if the biologist desires a different set of components then there is a need to the API so that their preferred components can be integrated. Finally, we thought by building the platform to run on desktop and laptop clients, rather than central servers, independent PIs would find it possible to configure the system with components for their personal use. However, most biological researchers with access to the web are more comfortable with a single web browser technology than multiple client side technologies.

The project has brought us a number of collaborative opportunities. In addition to the collaborations with the CGIAR, we have partnered with Cold Spring Harbor Laboratory on an NSF-funded project, and in cooperation with Canadian funding efforts, we are pursuing the design and implementation of a web-based, distributed analysis system. This project, called MOBY, is aimed at developing a common syntax, common semantic, and discovery mechanism for bioinformatic data and services. MOBY rests heavily on the ideas and approaches of ISYS and DAS (Dowell *et al.*, 2001). Conceptually, it is a small step to expand ISYS's loose-coupling, client-side approach to full web integration: that is, the association of web services with data types and service discovery. Architecturally, though, such web integration is non-trivial, as a number of issues concerning syntax, semantics, and discovery need to be addressed in the creation of a viable model.

## 6. SUMMARY OF LESSONS LEARNED FROM ALL INTEGRATION PROJECTS

In this manuscript I've used three criteria to judge the integration projects: financial considerations, scientific and technical innovation and use of the system for research and discovery by biologists. Evaluating the first criterion is straight forward with an accurate accounting system. Scientific and technical innovations are a little more difficult to measure, but certainly peer reviewed publications represent a measure that can be quantified. The most difficult to measure is system use. It is a criterion that software developers use to judge themselves, but it is difficult to tell if a system is being used by the number web-site visits, number of downloads or requests for licenses.

As a non-profit organization, where 100% of our annual operating expenses need to be covered by the R&D projects, we are learning that there are few funding opportunities from federal granting agencies for pure software development projects. Thus, we have broadened our funding portfolio to include research foundations and international agencies. XGI is our only project where virtually all direct labor and infrastructure costs have been fully covered by external funding; primarily through competitive grants. We chose to subsidize the remaining projects with internal funds for various reasons. We deliberately decided to subsidize ISYS because we wanted to pursue its scientific and technical innovations. Because TAIR is a high profile, long-term project we have been willing to subsidize unforeseen costs due to emerging requirements. Although these emerging requirements could have been better forecast in the original proposal, we have to recognize that projecting five years in a technically changing environment involves a certain amount of luck. GeneX has been our most ambitious project with regard to planning a system in a changing scientific and technical environment, but we also have to admit that a great deal of its initial costs can be attributed to poor management.

The project with the least innovative computer science and information technology, TAIR, has been the most successful in terms of serving the biological research community. On the other hand, our most innovative integration project, ISYS, has had the smallest impact in terms of serving the biological research community. Design and requirements of the former have been driven by biologists, while biologists were used primarily for "use-cases" in the development of the latter. This dichotomy between development needs of biologists and research interests of bioinformaticists is fairly widely recognized and represents a significant challenge for program directors at funding agencies. One way to address this is to utilize different mechanisms for development than for research, e.g., the recently issued RFA from NIAID for Bioinformatics Resource Centers. As Bioinformatics emerges as a discipline, however, it is likely that both research and development can and will be accommodated in large programmatic grants.

## 7. REFERENCES

Benton, D., 2000, Standards to Enable Bioinformatics Data and Information Integration, In *Barnett International's 2nd Annual Bioinformatics and Data Integration Conference*, Philadelphia, PA.

Boyle, J., 1998, Building Component Software for the Biological Sciences, *CCP11 Newsletter*, 4:22–14.

Dowell, R., Jokerst, A., Day, S., Eddy, L., and Stein, L., 2001, The distributed annotation system, *BMC Bioinformatics* **2**(7). This article is available at http://www.biomedcentral.com/1471-2105/2/7.

Fischer, S., Crabtree, J., Brunk, B., Gibson, M., and Overton, G.C., 1999, bioWidgets: data interaction components for genomics, *Bioinformatics*, 15:837–846.

Flanders, D.J.,Weng, S., Petel, F.X., and Cherry, J.M., 1998, AtDB, the Arabidopsis thaliana database, and graphical-web-display of progress by the Arabidopsis Genome Initiative. *Nucleic Acids Res* 26:80.

Gessler, D., 2002, ISYS: A Platform for Integrating Heterogenous Bioinformatic Resources. Comp Funct Genom 3:169–175.

Goodman, N., Rozen, S., and Stein, L., 1995, The Importance of Standards and Componentry in Meeting the Genome Informatics Challenges of the Next Five Years, In *Second Meeting on the Interconnection of Molecular Biology Databases*.

Huala, E., Dickerman, A.W., Garcia-Hernandez, M., Weems, D., Reiser, L., LaFond, F., Hanley, D., Kiphart, D., Zhuang, M., and Huang, W., 2001, The Arabidopsis Information Resource (TAIR): a comprehensive database and web-based information retrieval, analysis, and visualization system for a model plant, *Nucleic Acids Res*, 29, 102–105.

Karp, P., 1996, A strategy for database interoperation, *Journal of Computational Biology*, 2:573–586.

Kim, S., Lund, J., Kiraly, M., Jiang, M., Stuart, J., Eizinger, A., Wylie, B., and Davidson, G., 2001, A gene expression map for *Caenorhabditis elegans*, *Science* 293:2087–2092.

Matthews, W., Atlas, S.R., and Ruby. S.W., 2002, Visualizing genomic data according to chromosomal position, *Yeast Genetics and Molecular Biology Meeting*, Univ Wisconsin, Madison.

Object Management Group, 1996, *CORBA: Architecture and Specification*, OMG publication.

Ritter, O., 1994, The Integrated Genomic Database (IGD), In *Computational Methods in Genome Research*, S. Suhai, ed., Plenum Press, New York, pp. 57–73.

Rojas, P., 2000, Letting Go: Open Source Isn't for Everyone, Red Herring:296–306.

Searls, D., 1995, bioTK: Componentry for genome informatics graphical user interfaces, *Gene* 163, GCI-16.

Schena, M., Shalon, D., Davis, R.W., and Brown, P.O., 1995, Quantitative monitoring of gene expression patterns with a complementary DNA microarray, *Science* 270:467–470.

Siepel, A., Farmer, A., Tolopko, A., Zhuang, M., Mendes, P., Beavis, W.D., and Sobral, B.W.S., 2001a, ISYS: A decentralized, component-based approach to the integration of heterogeneous bioinformatics resources, *Bioinformatics* 17:83–94.

Siepel, A., Tolopko, A., Farmer, A., Steadman, P., Schilkey, F., Perry, B.D., and Beavis, W.D., 2001b, A development platform for the integration of heterogeneous bioinformatics software components, *IBM Systems Journal* 540–592.

Slidel, T., 1998, The Life Sciences Research Task Force. In *Objects in Bioinformatics '98*, Cambridge, England.

Szyperski, C., 1998, *Component Software: Beyond Object-Oriented Programming*, Addison Wesley Longman Limited, Essex, England.

Toner, B., 2002, For Bioinformatics Companies, 2002 Was the Year of 'Change or Die': What's next? *BioInform* 12:23–02.

Velculescu, V.E., Zhang, L., Vogelstein, B., and Kinzler, K.W., 1995, Serial analysis of gene expression, *Science* 270:484–487.

Vos, P., Hogers, R., Bleeker, M., Reijans, M., Lee, T.V.D., Hornes, M., Frijters, A., Pot, J., Peleman, J., Kuiper, M., and Zabeau., M., 1995, AFLP: a new technique for DNA fingerprinting, *Nucleic Acids Res.* 27:1.

*Chapter 4*

# Functional Genomics Approach to Elucidate the Regulation of Vascular Development in Poplar

Rishikesh P. Bhalerao and Göran Sandberg

## 1. INTRODUCTION

The vascular cambium is the lateral meristem responsible for wood formation in trees (Larson 1994). The cambial initial (equivalent of the stem cells) undergoes cell divisions that give rise to xylem and phloem mother cells. These mother cells undergo several rounds of cell division to give rise to daughter cells that progressively undergo expansion, secondary wall formation and cell death. This sequential progression of the cells from division to death manifests itself in an easily distinguishable developmental gradient on the xylem side (Figure 1). The regular pattern of wood formation is altered by environmental and hormonal signals. For example, wood formation differs noticeably during the growing season with xylem cells being thin walled early in season where as they are thick walled during the later part of the season (Larson 1960). Similarly, bending trees modulates the orientation of the cambial cell divisions, with the cambial cell divisions terminating on one side and being enhanced on the opposite side. Additionally, the walls of these newly formed xylem cells have higher cellulose to lignin ratio (Timell 1969). Age of the tree also influences wood formation with juvenile woodbeing different compared to mature wood. Finally, one of the most significant alterations in wood formation

**Rishikesh P. Bhalerao and Göran Sandberg**   Umeå Plant Science Center, S-901 87, Umeå, Sweden

*Genome Exploitation: Data Mining the Genome*, edited by J. Perry Gustafson, Randy Shoemaker, and John W. Snape.
Springer Science + Business Media, New York, 2005.

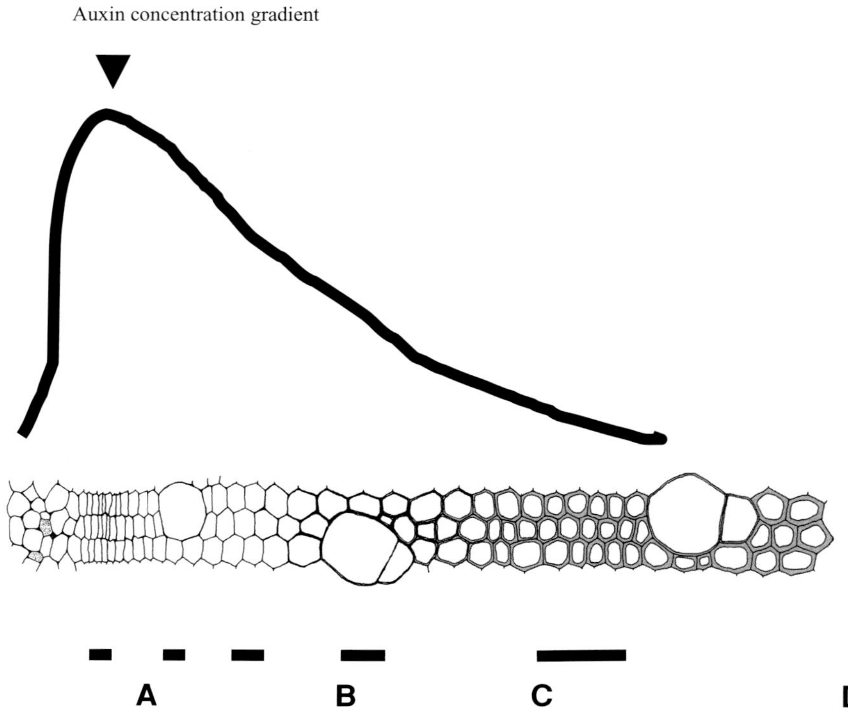

Auxin concentration gradient

A          B          C          D

**FIGURE 1.** A schematic representation of the wood forming zone in trees with auxin concentration gradient overlapping the developmental gradient in the secondary xylem tissues. The main zones shown are, A-dividing cambial cells, B-early expanding cells, C-late expanding cells, D-early secondary wall forming cells and E-late secondary wall forming cells. Please note that this zone also includes cells undergoing cell death.

is the seasonal change in cambial activity so that the period of wood formation is synchronized with the period of favourable growth conditions. This cycling of the cambium between active and dormant state allows the trees to protect the meristems from harsh environmental conditions as well (Little and Bonga 1974).

Thus from above it is clear that wood formation is influenced by several different environmental and other signals. The challenge for tree biologists is to understand the molecular basis of wood formation and its alteration by a diverse range of signals. Understanding of wood formation is important from both research and economical point of view. On one hand wood formation provides an excellent experimental system to investigate the developmental regulation. The spatial separation of xylems cells at distinct stages of development permits their isolation for high resolution transcript profiling and growth regulator measurements (Uggla et al., 1996; Hertzberg et al., 2001). At the same time wood production is economically important. However, the investigation of wood formation especially at the molecular level has significantly lagged behind other processes for several reasons until recent times. This paper evaluates the progress made in the area of molecular wood biology with emphasis on the work that has been performed in poplar, which has become an experimental model for elucidating wood formation using functional genomics approach. In this chapter we will focus on two areas of wood formation, (i) The development of secondary xylem and (ii) The seasonal control of cambial activity.

## 2. POPLAR AS A MODEL FOR STUDYING FORMATION

One of the major reasons for the lack of knowledge regarding wood formation especially at the molecular level has been a lack of good experimental systems to investigate wood formation. While trees such as pine, spruce are economical important, the long generation times and that these trees are not easily amenable to genetic approaches has precluded their use for the purpose for experimental analysis of wood formation using molecular techniques with few exceptions (MacKay et al., 1997; Allona et al., 1998). Equally important is the fact that wood as a tissue is difficult to obtain for experimental manipulation although at the same time the basic design of wood cells make them an excellent experimental system. For example, the large size of cambial meristem and the fairly regular pattern of cells in the secondary xylem allows measurement of the concentration of key growth regulators such auxin and transcript pools of xylem cells at specific stages of development. This in turn allows prediction of the influence of concentration gradients of growth regulators and alterations in the pattern of gene expression to build a transcriptional network to explain the molecular basis of xylem development. Several years ago scientists in Umeå and other groups in several other labs focused on poplar as an experimental system for questions pertaining to tree biology (Feuillet et al., 1995; Tuominen et al., 1995; Nilsson et al., 1996). This choice of poplar as an experimental system was based the rapid growth rate of polar and the ability to manipulate poplar genetically through agrobacterium mediated transformation. In retrospect, this choice of poplar was especially fortuitous given that DOE decided to sequence poplar genome in 2001 and completed its sequence in December 2002 (http://genome.jgi-psf.org/poplar0/poplar0.home.html).

## 3. ELUCIDATING THE CONTROL OF SECONDARY XYLEM DEVELOPMENT

### 3.1. The Role of Aux/IAA Genes in Regulation of Secondary Xylem Development

One of the key findings in the area of vascular development was the description of a concentration gradient of plant growth regulator auxin overlapping the developmental gradient of xylem in the wood forming tissues in trees (Uggla et al., 1996) (Figure 1). The concentration of auxin displays a peak in zone comprising the

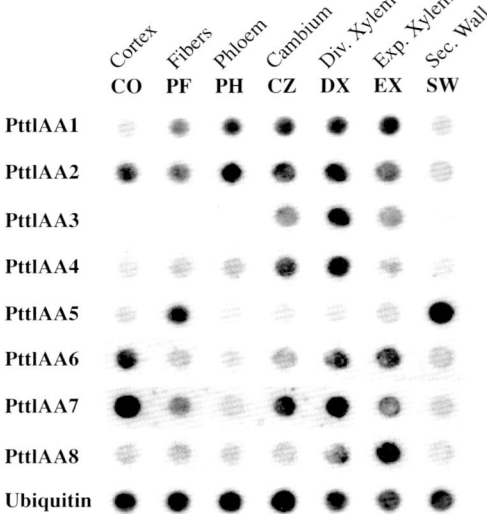

**FIGURE 2.** Expression pattern of hybrid aspen Aux/IAA genes (denoted as PtIAA) in the cell types comprising the wood forming zone. Note that cambium corresponds to zone A, div xylem (dividing xylem) corresponds to zone B, expanding xylem (Exp Xylem) corresponds to zone C and secondary wall forming xylem (Sec xylem) corresponds to zone D of figure 1.

cambial initial and dividing cells and then progressively declines towards the edge of xylem in the zone comprising the secondary wall forming cells. Based on this distribution of auxin it has been proposed that auxin concentration acts as a positional cue to influence xylem development. This hypothesis leads to the question of how the concentration gradient of auxin can translate into activation and repression of gene expressions whose output is the visible pattern of xylem development.

The answer to these questions has been approached in two ways. Firstly we have cloned cDNAs for 18 members of the Aux/IAA gene family from poplar. The Aux/IAA genes encode small molecular weight nuclear targeted proteins that interact with ARF family of transcription factors and influence gene expression in auxin responsive manner (Ouellet et al., 2001; Tiwari et al., 2001). Important to this discussion is also the observation that Aux/IAA proteins are destabilized by auxin (Gray et al., 2001; Tiwari et al., 2001). Furthermore, mutations in several Aux/IAA genes lead to aberrant auxin responses in Arabidopsis (reviewed in Reed 2001). In order to elucidate the regulation of xylem development by auxin we examined the expression of poplar Aux/IAA genes in the cell types comprising the developmental gradient of secondary xylem (Moyle et al., 2002). The results shown in figure 2 indicate the expression of Aux/IAA genes is highly specific expression patterns. The pattern of expression of the poplar Aux/IAA genes suggest two potential mechanisms how auxin distribution can influence the course of xylem development. Firstly, the different Aux/IAA genes can act as tissue specific auxin responsive transcription factors that activate or repress the expression of downstream genes. These auxin dependent activation activation or repression of gene expression programs could then modulate the course of xylem development leading to organization of developing xylem cells into distinct developmental zones. An alternative explanation of auxin regulation of secondary xylem development can be considered based on the observation that auxin concentration can regulate the level of Aux/IAA proteins. This would mean that auxin concentration gradient could set a corresponding gradient of Aux/IAA transcription factors. This gradient of Aux/IAA transcription factors in turn would activate or repress downstream genes depending upon the concentration of the Aux/IAA transcription factor leading to auxin dependent activation or repression of gene expression programs in the developing secondary xylem cells.

## 3.2. Genomics Approach to Elucidation of Secondary Xylem Development

### 3.2.1. Large Scale Gene Discovery in Wood Forming Zone of Poplar

The auxin concentration gradient and the gene expression of Aux/IAA genes in the xylem cells provides the conceptual framework that at least partly explains how auxin concentration gradient may regulate the formation of distinct domains of developing xylem cells. However, the downstream targets of these Aux/IAA genes need to be described, a task that cannot be accomplished by conventional approaches. To illustrate this point it is useful to consider that at the start of the poplar project (described below) in 1998, less then 50 cDNA sequences were present in the NCBI database. Therefore the first step towards understanding the xylem development was to initiate gene discovery in poplar. As realised in several other organisms, the most obvious way to do this was to perform EST sequencing in poplar preferably using cDNA library prepared from RNA isolated from the tissue of interest, the developing xylem. Therefore an EST sequencing project was initiated and 5000 ESTs were sequenced from a cDNA library constructed from mRNA isolated from developing xylem and phloem tissues (Sterky et al., 1998). The ESTs sequenced provided the first glimpse of the type genes that are expressed in wood forming cells. A total of 2,988 transcripts could be identified from the assembly of sequenced ESTs. Several genes for lignin and cellulose biosynthesis were identified as well. For almost 40% of the sequenced ESTs, no function could be deduced from the sequence alone. However it is important to realize that this could be due to either small sequence length and/or the fact that 5′ sequence does not contain enough information for confirming sequencing identity.

### 3.2.2. Transcript Profiling of Developing Xylem

While EST sequencing was useful in describing the type of genes expressed in wood forming cells, these results do not provide information on how different gene expression programs are regulated during the course of xylem development. Therefore transcript profiling of developing xylem was performed. In order to perform transcript profiling a microarray was fabricated. The sequenced ESTs

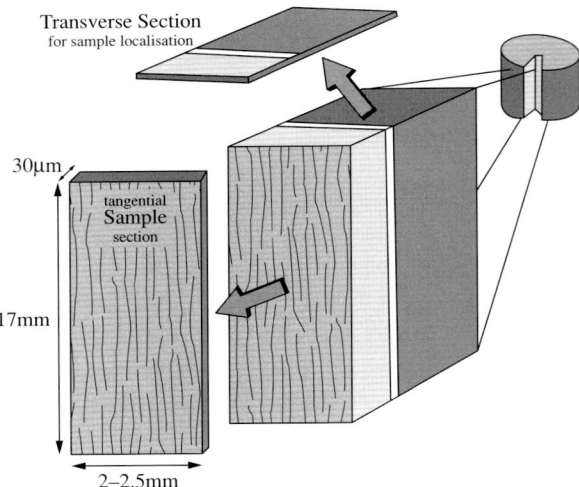

**FIGURE 3.** Cryosectioning technique to obtain 1–3 cell layer thick sections for mRNA isolation. See section 4.2.1 for description.

were assembled creating a unigene set of approximately 2300 unique sequences. The ESTs for these 2300 unique "genes" were used to generate a cDNA microarray for transcript profiling of cells comprising the developmental gradient of the xylem cells (Hertzberg et al., 2001).

In order to investigate the choreography of gene expression to chart the course of xylem development, it was necessary to perform transcript profiling using probes prepared from single cell layers. In order to do this, firstly, we developed cryosectioning technique to obtain tangential sections of developing xylem cells (Uggla et al., 1996) (see Figure 3). This technique allowed precise isolation of well defined 1–3 cell layers of xylem about 30 micrometer in size. The sections isolated in this manner were used to measure auxin distribution across the developing xylem cells. Typically, it was calculated that the tissue obtained using this method would be about 0.5 milligrams and therefore a yield of 0.5 micrograms of RNA was expected from such a small amount of sample. The challenge was in being able to isolate RNA from this tissue and make probes using extremely small amount of RNA. This difficulty was due to the existing probe labeling protocols that use at least 1–2 micrograms of mRNA and therefore a novel PCR amplification based protocol was developed (Hertzberg et al., 2001). Briefly, the method was based on random fragmentation of cDNA obtained from small amounts of tissues obtained by cryosectioning (typically 1–3 cell layers of 30 micrometer) followed by capture of the 3 prime ends using magnetic beads technology followed by PCR amplification of the 200–400 base pair long fragments to generate the target that could be labeled for use in transcript profiling experiments. This method overcame the problems associated with alternative PCR based methods using mRNA isolated from single cells which suffered from the bias in PCR towards small fragments which was further reflected in results obtained in transcript profiling experiments. The hybridization results from the amplification strategy were compared with the results obtained using conventional probe preparation methods. It was found that the amplification method was robust and reproducible.

### 3.2.3. The Spatial Regulation of Gene Expression During the Secondary Xylem Development

In order to describe the dynamics of the pattern of gene expression during xylem development, transcript profiling was performed (Hertzberg et al., 2001). For transcript profiling, probes were generated from RNA isolated from 30 micrometer sections as described above. In total 5 sections were used referred to as ABCDE in the subsequent figures. A corresponded to cambium, B-young xylem, C-expanding xylem, D-secondary wall forming xylem and E-the zone of programmed cell death. The experimental design for transcript profiling involved hybridizing each zone against a pooled reference sample i.e. A/ABCDE and B/ABCDE, this allowed a direct comparison of A vs B. The results of this analysis are summarized below and explain some of the information obtained from this analysis.

Firstly it was found that although the zone of cell division is confined to zone A, the expression of cell cycle genes e.g. *PttCDKA*, *PttCYCH* extends beyond this zone. This may reflect that the cells retain their competence to divide late into development. Also this may be indicative of the fact that some of the cell cycle genes encode proteins that may play a role in other aspects of cellular functions. Secondly, our microarray analysis was useful in delineating the potential roles for individual members of the large gene families. For example, we found over 16 sequences for tubulins and their expression patterns could be divided into 2 clusters. One of these had a peak of expression in the zone B where most of the dividing cells are present where as the other cluster was positioned over the zone of expansion. Tubulins are involved in cell division as well as orientating the cellulose microfibirils and these two processes are separated in space and the distinct expression of tubulins may indicate the roles of the different members of the tubulin family. Similarly, the members of MYB transcription family also exhibited differential expression across the cells of developing xylem. As MYB genes have been involved in regulation of diverse cellular processes ranging from cell division to lignification, the description of the expression patterns of different members of this family provides the first clue regarding the potential roles in the context of xylem development. These examples simply reflect the type of information obtained from global transcript profiling. Thus the data from microarray analysis not only described the genes that are expressed in specific cell types but more importantly they allowed us to construct the transcriptional network underlying the formation and organization of the domains of the cells comprising the developing xylem for the first time.

## 4. THE SEASONAL CONTROL OF CAMBIAL ACTIVITY

The other key aspect of vascular development considered here is the seasonal cycle of the vascular cambium. One of the major points of distinction between annuals such as Arabidopsis and perennials such as trees is the seasonal cycle of the activity of the cambial meristem. The cambial meristem cycles between active and dormant states. This cycling allows the cambial meristem to synchronise its period of growth with favourable environmental conditions thereby maximizing its period of growth and protecting the meristem from harsh environmental conditions (reviewed in Samish 1954). However the activity-dormancy cycling is a complex process that involves several stages (see Figure 4) whose orderly transition is regulated by the environmental and possibly hormonal signals. The cycling of the cambial meristem between active and dormant state is described in the figure 4. The cambium during the active period of growth can be considered to be in online state with the cell division machinery being responsive to growth promotive signals. However upon exposure of the trees to short days, growth cessation occurs and cambium enters the ecodormant state at this stage the cell division machinery can be considered to be in standby mode as transfer of the plants in ecodormant state to favorable conditions leads to reactivation of the growth. Later in the season, the cambium makes the transition to endodormant state with the cell division machinery entering the offline mode, becoming insensitive to the growth promotive signals. Exposure to chilling temperatures leads to transition of the cambium into ecodormant state and the cambial cell division machinery returns to standby

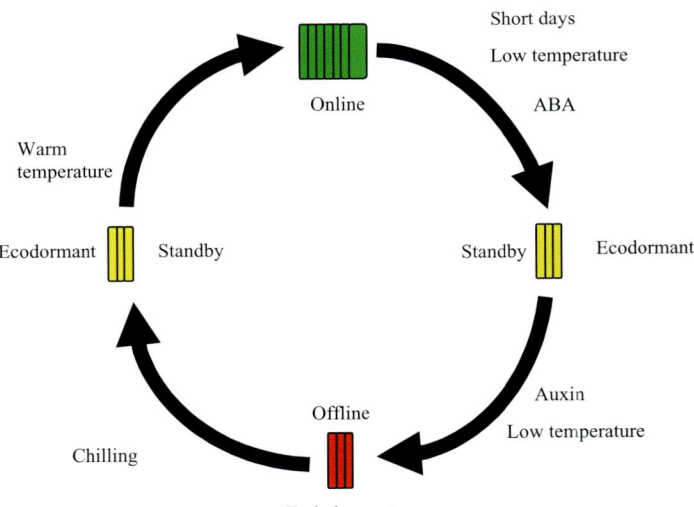

**FIGURE 4.** Schematic representation of the annual cambial cycle. See the description for explanation of terms.

mode so that warm spring temperatures cause reactivation of the cambial growth. At this stage very little is known about the molecular basis of regulation of the cambial activity by environmental signals and in particular about the differences between ecodormant and endodormant states at the molecular level. The entire process of cambial cycle involves several sub processes, in which multiple signals, environmental and hormonal act on diverse signaling pathways. Therefore it was deemed that the study of the regulation of cambial activity would benefit from a functional genetics approach. However, prior to initiating a functional genomics approach to study cambial cycle it is important to remember that the actual site of dormancy is the cambial meristem. Therefore obtaining relevant information regarding dormancy would benefit considerably from being able to analyse prior to initiating a functional genomics approach to study cambial cycle it is important to remember that the actual site of dormancy is the cambial meristem. Therefore obtaining relevant information regarding dormancy would benefit considerably from being able to analyse gene expression in the cambium itself. This is especially a relevant issue since the meristem constitutes a minor part of stem or apical bud and therefore it is to be expected that grinding up entire stem or apical bud might cause considerable loss of information due dilution effects. Importantly, it is the meristem that undergoes dormancy and not the surrounding tissues. Therefore in our investigations of cambial activity we worked with cambial sections for making cDNA libraries as well as transcript profiling as described below. We used cryosectioning described earlier to isolate pure cambial sections as a starting material for cDNA library construction as well as transcript profiling experiments. This minimized confounding of results from processes that occur in non-meristematic tissues during dormancy in the cambium.

## 4.1. Large Scale Sequencing of Active and Dormant Cambial Libraries

In order to analyse seasonal cambial activity we firstly generated cDNA libraries for active and dormant cambium and sequenced about 4–4.5 thousand ESTs to identify the genes expressed in active and dormant cambium and compare the difference in expression between active and dormant cambium to get a first glimpse of the changes in gene expression taking place upon transition from active to dormant state. The data indicated that less then 10% of ESTs were common between the two libraries. While it is clear that the total number of ESTs sequenced are not high, nevertheless, this data clearly indicates that active to dormant transition involves considerable change in cambial gene expression. Secondly, the EST assembly of active and dormant libraries independently leads to a greater number of assembled genes from active cambial library compared to dormant library indicating that there is greater diversity of gene expression in the active cambium compared to that in the dormant cambium.

One of the interesting observations of EST sequencing of cambial cDNA library was that the pattern of meristematic transcriptome clearly differs from the transcriptome of developing leaf. This is most clearly reflected in the difference in the percentage of ESTs involved in energy generation. In case of the leaf cDNA library, this number is in the range of 36% where as for cambium it is in the range of 5–6%. The other interesting aspect of active and dormant cDNA library comparison indicates that there is little quantitative difference in the percentage of metabolism and energy subclasses between cambium. However qualitatively, secondary metabolism is considerably increased in the dormant cambium, which is further supported by the increased expression of lignan and terpenoid metabolism genes in the dormant cambium compared to active cambium.

## 4.2. Transcript Profiling of Active and Dormant Cambium

While EST sequencing provided us with important information on the transcriptome of a meristematice tissue, it was important to analyse not only the type of genes that are expressed in active and dormant cambium but their relative levels in cambium at two different states. Therefore we performed global transcript profiling to compare the expression of cambially-expressed genes in active and dormant state. The data is too numerous to describe in entirety here but some major observations are described below. Firstly, we found about 4000 genes that were significantly upregulated in the active cambium compared to dormant cambium and about 700 genes that were upregulated in the dormant cambium compared to active cambium. Here the results from dormant cambial gene patterns are described. In accordance with cambial insensitivity to auxin during endodormancy, one of the genes SINA that is a negative regulator of auxin sensitivity (Xie et al., 2002) was upregulated in the dormant cambium. Similarly, genes are involved in starvation responses such as poplar homolog of SIP4. Analysis of the transcript profiling data is time consuming and therefore a methodology needs to be developed that would reduce the time required to analyse this data. We therefore concentrated our analysis on looking into genes that constitute an entire pathway e.g. glycolysis etc. to identify pathways rather then genes that are potentially upregulated during dormancy. This is exemplified by the two genes encoding malate synthase and isocitrate lyase that often are coregulated and are involved in glyoxalate cycle channeling lipids for use as energy and/or carbon source (Rylott et al., 2001). Interestingly, these genes are often upregulated during seed germination to drive the initial phase of growth before photosynthesis can provde the plants with energy. This observation leads to a question whether the transcripts for the abovementioned genes are also translated or whether they are stored to be translated during the initial phase of cambial reactivation before newly formed leaves can photosynthesis and provide energy for cambial growth. In addition to these genes other genes upregulated in the dormant cambium includes genes encoding enzymes of lipid synthesis, sugar metabolism etc. This data is a mere tip of the iceberg and considerably more bioinformatics analysis is necessary to glean useful information for the future. The obvious next step will be to perform a detailed time course of cambial gene expression as cambium makes the transition from active to dormant state to improve upon the results here.

# 5. FUTURE PERSPECTIVES

## 5.1. Arabidopsis as Test Bed for Rapidly Identifying Genes Involved in Vascular Development

Poplar is an excellent experimental system to perform transcript profiling and metabolic profiling in the vascular cells owing to the large size of the cambial region to identify important processes and genes whose expression suggests a role in regulating vascular development. However, the subsequent genetic analysis is time and resource consuming. In particular it is important to note that several of genomics technologies rapidly identify genes of interest but this itself creates a functional genomics nightmare since because of the sheer amount of information, it becomes difficult to glean useful information. Typically, there are several genes whose function cannot be predicted based on sequence alone. Secondly, there are multigene families and it is difficult to understand the role of individual genes in a specific process. Thirdly, there are genes for which there is no closely related sequence in the database yet. In view of these difficulties, it is necessary to identify a model system in which functional analysis of genes of interest identified from experiments in poplar can be performed with ease. The model plant Arabidopsis is probably the best experimental system for analysis of plant gene function at the moment. With the entire genome sequenced and several mutant collections existing makes it fairly easy to understand the function of the poplar gene for which a gene with high sequence identity in Arabidopsis can be found (see www.arabidopsis.org for the description of all the resources). However, in case of vascular development it was important to first characterize the vascular development in Arabidopsis and ascertain that it is similar to that in poplar. As can be seen in figure 5, Arabidopsis is able to undergo secondary growth and major developmental events and the components of the vascular system show high degree of similarity with poplar (Chaffey et al., 2002). This high degree of conservation of secondary growth and vascular development between poplar and Arabidopsis meant that we could use Arabidopsis as test bed for rapidly characterizing the function of genes from poplar whose sequence does not predict their function but whose expression pattern links them to a specific process in vascular development. Below we give examples where this strategy has been tested.

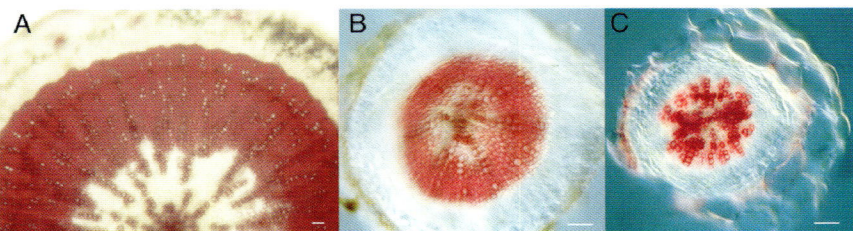

**FIGURE 5.** Transverse section of Arabidopsis hypocotyls showing secondary growth (A and B) Wild type Arabidopsis, (C) A mutant with inactivated gene of GRAS family

## 5.2. Rapid Analysis of Gene Function of Poplar Genes with Unknown Function

One of the key observations that arises from genomics approach for elucidating vascular development is that both EST sequencing as well gene expression analysis leads to identification of several genes whose function can not be predicted based on sequence or expression pattern alone. The number of genes that fall into this category is quite high. It is extremely essential to develop strategies to rapidly identify the function of these genes. We have taken an approach to rapidly understand the function of poplar genes using *Arabidopsis thaliana* as a model system. The strategy is based on identify Arabidopsis genes that display high level of sequence similarity to poplar genes of interest. Following this, in those cases where a single Arabidopsis gene with high homology to the poplar gene of interest is identified, in silico screening is performed to identify Arabidopsis mutants in which a T-DNA has been inserted thereby inactivating the function of the Arabidopsis gene. With several insertional knockout collections being available, the probability to find a T-DNA mutant is very high. This strategy was applied to 150 poplar genes that displayed a specific expression pattern in the zones A and B (Figure 1). Analysis of Arabidopsis genome sequence led to the identification of 15 poplar genes for which either 1 or maximum of 3 Arabidopsis genes with high sequence homology could be identified. The sequence of these Arabidopsis genes was used to screen the publicly available T-DNA mutant collections and in all cases at least 1 T-DNA tagged Arabidopsis mutant was identified (A. Marchant, pers communication). Of the 15 mutants so analysed, 2 displayed a clear visual phenotype. In both the cases, the mutants had apparent alterations in cell elongation (R. Nilsson pers communication). Currently we are performing analysis of their vascular development. In another case, a transcription factor of GRAS family (Pysh et al., 1999) whose expression was limited to cambial cells was used to identify closely related genes in Arabidopsis. This led to identification of over 8 Arabidopsis genes that displayed high level of similarity to poplar genes. Since characterization of the mutants of all Arabidopsis GRAS-like genes would be time consuming, we proceeded to characterize the expression of arabidopsis genes with vascular expression. One of the 8 genes had expression in the vascular tissue. A T-DNA tagged mutant for this gene was obtained and it displayed aberrant vascular development (Figure 5) indicating the promise of the strategy outlined above.

## 6. CONCLUSIONS

To summarize, poplar appears to be a very useful experimental model to analyse wood formation at the molecular level. With large EST collections and full genome sequence publicly available, genomics approach, which has been successfully used for elucidating the regulation of Arabidopsis can be used in poplar to a great extent. Finally, the conservation at the genetic and developmental level between poplar and Arabidopsis at least in terms of fundamental processes related to wood formation promises the possibility to use Arabidopsis as a test bed for analysis of the function of genes involved in wood formation.

ACKNOWLEDGEMENTS. The authors wish to thank all the members of UPSC and Royal Institute of Technology associated with poplar genomics program for fruitful collaborations that constitute much of the data described in this paper. The authors wish to thank the Wallenberg Consortium for funding the Poplar Genomics program in Sweden. The work in the lab of Dr. Bhalerao is funded by STEM and VR grants and in the lab of Prof. Sandberg by VR and WCN.

# 7. REFERENCES

Allona, I., Quinn, M., Shoop, E., Swope, K., St Cyr, S., Carlis, J., Riedl, J., Retzel, E., Campbell, M.M., Sederoff, R., and Whetten, R.W., 1998, Analysis of xylem formation in pine by cDNA sequencing, *Proc Natl Acad Sci U S A*. **95**: 9693.

Chaffey, N., Cholewa, E., Regan, S., and Sundberg, B., 2002, Secondary xylem development in Arabidopsis: a model for wood formation, *Physiol Plant*. **114**: 594.

Feuillet, C., Lauvergeat, V., Deswarte, C., Pilate, G., Boudet, A., and Grima-Pettenati, J., 1995, Tissue- and cell-specific expression of a cinnamyl alcohol dehydrogenase promoter in transgenic poplar plants. *Plant Mol Biol*. **27**: 651.

Gray, W.M., Kepinski, S., Rouse, D., Leyser, O., and Estelle, M., 2001, Auxin regulates SCF(TIR1)-dependent degradation of AUX/IAA proteins, *Nature*. **414**: 271.

Hertzberg, M., Aspeborg, H., Schrader, J., Andersson, A., Erlandsson, R., Blomqvist, K., Bhalerao, R., Uhlen, M., Teeri, T.T., Lundeberg, J., Sundberg, B., Nilsson, P., and Sandberg, G., 2001, A transcriptional roadmap to wood formation, *Proc Natl Acad Sci U S A*. **98**: 14732.

Hertzberg, M., Sievertzon, M., Aspeborg, H., Nilsson, P., Sandberg, G., and Lundeberg, J., 2001, cDNA microarray analysis of small plant tissue samples using a cDNA tag target amplification protocol, *Plant J*. **25**: 585.

MacKay, J.J., O'Malley, D.M., Presnell, T., Booker, F.L., Campbell, M.M., Whetten, R.W., and Sederoff, R.R., 1997, Inheritance, gene expression, and lignin characterization in a mutant pine deficient in cinnamyl alcohol dehydrogenase, *Proc Natl Acad Sci U S A*. **94**: 8255.

Moyle, R., Schrader, J., Stenberg, A., Olsson, O., Saxena, S., Sandberg, G., and Bhalerao, R.P., 2002, Environmental and auxin regulation of wood formation involves members of the Aux/IAA gene family in hybrid aspen, *Plant J*. **31**: 675.

Larson, P.R., 1960, A physiological consideration of the springwood summerwood transition in red pine, *Forest Sci*, **6**: 110.

Larsson, P.R., 1994, *The Vascular Cambium Development and Structure*, Springer-Verlag, Berlin.

Nilsson, O., Little, C.H., Sandberg, G., and Olsson, O., 1996, Expression of two heterologous promoters, Agrobacterium rhizogenes rolC and cauliflower mosaic virus 35S, in the stem of transgenic hybrid aspen plants during the annual cycle of growth and dormancy, *Plant Mol Biol*. **31**: 887.

Ouellet, F., Overvoorde, P.J., and Theologis, A., 2001, IAA17/AXR3: biochemical insight into an auxin mutant phenotype, *Plant Cell*. 13: 829.

Pysh, L.D., Wysocka-Diller, J.W., Camilleri, C., Bouchez, D., and Benfey, P.N., 1999, The GRAS gene family in Arabidopsis: sequence characterization and basic expression analysis of the SCARECROW-LIKE genes, *Plant J*. **18**: 111.

Reed, J., 2001, Role and activities of Aux/IAA proteins in Arabidopsis, *Trends in Plant Sci*. **6**: 420.

Rylott, E.L., Hooks, M.A., and Graham, I.A., 2001, Co-ordinate regulation of genes involved in storage lipid mobilization in Arabidopsis thaliana, *Biochem Soc Trans*. **29**: 283.

Sterky, F., Regan, S., Karlsson, J., Hertzberg, M., Rohde, A., Holmberg, A., Amini, B., Bhalerao, R., Larsson, M., Villarroel, R., Van Montagu, M., Sandberg, G., Olsson, O., Teeri, T.T., Boerjan, W., Gustafsson, P., Uhlen, M., Sundberg, B., and Lundeberg, J., 1998, Gene discovery in the wood-forming tissues of poplar: analysis of 5, 692 expressed sequence tags, *Proc Natl Acad Sci U S A.* **95**: 13330.

Timell, T.E., 1969, The chemical composition of tension wood, *Svensk papperstidning.* **72**: 173.

Tiwari, S.B., Wang, X.J., Hagen, G., and Guilfoyle, T.J., 2001, AUX/IAA proteins are active repressors, and their stability and activity are modulated by auxin, *Plant Cell.* **13**: 2809.

Tuominen, H., Sitbon, F., Jacobsson, C., Sandberg, G., Olsson, O., and Sundberg, B., 1995, Altered Growth and Wood Characteristics in Transgenic Hybrid Aspen Expressing Agrobacterium tumefaciens T-DNA Indoleacetic Acid-Biosynthetic Genes, Plant Physiol. **109**: 1179.

Ugzla, C., Moritz, T., Sandberg, G., and Sundberg, B., 1996, Auxin as a positional signal in pattern formation in plants, *Proc Natl Acad Sci U S A.* **93**: 9282.

Xie, Q., Guo, H.S., Dallman, G., Fang, S., Weissman, A.M., and Chua, N.H., 2002, SINAT5 promotes ubiquitin-related degradation of NAC1 to attenuate auxin signals, *Nature.* **419**: 167.

*Chapter 5*

# Novel Tools for Plant Genome Annotation and Applications to *Arabidopsis* and Rice

Volker Brendel

## 1. INTRODUCTION

Our conference title, "Genome Exploitation: Data Mining", is equally evocative and provocative. The combined use of the terms "exploitation" and "mining" conjures up images of the industrial process of exploiting natural resources by mining the earth for precious metals. Such parallels are not accidental but reflect a widely accepted paradigm of modern genetics. This paradigm is evident in many other terms that have become common in our field, including bio*technology*, genetic *engineering*, *high-throughput* approaches, etc. It may be a bit too early to circumscribe this paradigm precisely. Undoubtedly, this will be done in the introductory chapters of forthcoming textbooks. But the dominant theme can be pinpointed as the industrialization of molecular biology and genetics. In this article I shall raise concerns about eager acceptance of this paradigm and argue for a more balanced approach. Personally, I would have preferred a conference title like "Genome Exploration: New Frontiers". Such title would place emphasis on discovery and learning, and it may suggest a more measured pace towards applications. In the technical part of this article I review my research group's efforts to provide

**Volker Brendel**    Department of Genetics, Development and Cell Biology and Department of Statistics, Iowa State University, Ames, Iowa 50011-3260

*Genome Exploitation: Data Mining the Genome*, edited by J. Perry Gustafson, Randy Shoemaker, and John W. Snape.
Springer Science + Business Media, New York, 2005.

web-accessible tools for plant genome annotation that attempt to harness individual, expert contributions to complement large-scale, "industrial" annotation efforts.

## 2. BOON AND BANE OF THE INDUSTRIALIZATION OF MOLECULAR BIOLOGY AND GENETICS

The Human Genome Initiative was an unprecedented large-scale international collaborative effort that resulted in the completion of the sequence of a representative human genome in 2001. To a large extent, this project was driven by medical research questions, both in the public and private sector. The National Plant Genomics Initiative in the United States (2003) and related programs in other countries have had a similar great impact on plant genome research, with *Arabidopsis thaliana* the first plant genome nearly finished three years ago (The Arabidopsis Genome Initiative, 2000). Again, much of this effort has been driven by practical applications. The promise of another "green revolution" by novel crop varieties has been attracting large and small businesses. Growers and the public have successfully lobbied for public projects to provide a counterpoint to proprietary research and ensure public control. The influx of funds and the development of new technologies have ushered a very exciting period in plant genome research, stepping at least partly out of the shadow of the much more publicized and funded medical research. However, the accelerated pace of research and data accumulation has also created problems. Our large-scale projects are generating such a volume of data that the responsible investigators cannot individually encompass all the data generated in their own projects. By contrast, think of the people who sequenced the first proteins, RNA, or DNA molecules—they would certainly have been able to write down these sequences from memory, residue by residue! In fact, the entire compilation of sequences fit into a nice little booklet only about 30 years ago. Now we can generate megabases of DNA sequence per day. Thus, automated annotation becomes necessary and quality control is a big issue. Of course, not only are we individually unable to keep pace with the sequence accumulation, but similar problems also occur with respect to the scientific literature. Thus, as our databases increase, instances of wrong annotation also increase, and these mis-annotations often propagate as new sequences get annotated based on similarity and implied annotation transitivity.

In the rush to generate more data, are we creating a backlog of data that is not sufficiently analyzed, annotated, and made easily accessible? Campbell and Karlin (2003) hint at this in their overview to a special issue of Theoretical Population Biology on the evolution of genome structures when they "pose the question of whether the primary need is for yet more data (how prioritized?) or for additional analyses of available databases." Konopka (2003) writes very critically about current genome annotation strategies and strongly advocates intellectual investment into new ideas and methods. The examples discussed below suggest that community efforts to correct and refine approximate, first-phase automated genome annotation may be a practical step to generate reliable data, which in turn are essential for building better theoretical models that may give rise to improved automated annotation tools. Building a more reliable foundation for further genome research not only serves our immediate research interests. It will also set a tone in our interactions with the public at large, increasing confidence that our knowledge base is solid and not compromised by the all too frequent fame-seeking public announcements of another finished genome (including the same one that was announced finished every other year before!) and more possibilities for lucrative biotech manipulations.

## 3. THE GENE STRUCTURE ANNOTATION PROBLEM

Complete genome annotation can be a very large problem. Initially, we think of this problem as identifying the genes in the genome, or in an even narrower sense, as identifying the protein coding regions. However, one quickly wishes for more: annotation of transcribed, but non-coding exons and parts of exons; identification of regulatory regions; indication of alternative splicing; levels of gene expression under different conditions; allelic variations; known mutants and their associated phenotypes; comparisons with orthologous loci and syntenic regions in other genomes. In this purview, the genome simply becomes the scaffold and portal for the entire body of associated genetic data, stored in linked databases. Because much of the other data rely on the gene structure annotation, accurate and comprehensive descriptions of gene structure are the central problem of genome annotation. In simple terms, the challenge may be posed as generating the mRNA and CDS tags of a standard GenBank file representing genomic DNA. The challenge exists both with respect to correcting existing annotation and with respect to (automated) generation of annotation for novel genomic sequence data. An "old" paper by Korning et al. (1996) illustrated typical problems in GenBank annotation of *Arabidopsis thaliana* gene structures. Unfortunately, the problem persists, and in particular, even the published annotation mistakes have mostly remained uncorrected (see Brendel, 2002). Because GenBank and other public databases function as repositories, incorrect annotations will be kept in these databases as historical, original records unless corrected by the submitting authors—an unlikely event in most cases. Thus, I only see two practical outs from the annotation problem: one, commercial interests will provide for a fee proprietary, curated databases based on public data but with corrected and enhanced annotation; second, research communities make a communal effort to re-annotate the genomes of their interest, with the load of human-intensive non-automated annotation distributed over the network of experts in the community. I shall discuss our attempts at promoting the latter approach, with *Arabidopsis thaliana* as the model.

## 4. GENE STRUCTURE ANNOTATION BY SPLICED ALIGNMENT

Gene structure prediction by *ab intio* approaches that rely on evaluation of genomic DNA sequence features to assign exon, intron, or intergenic status to

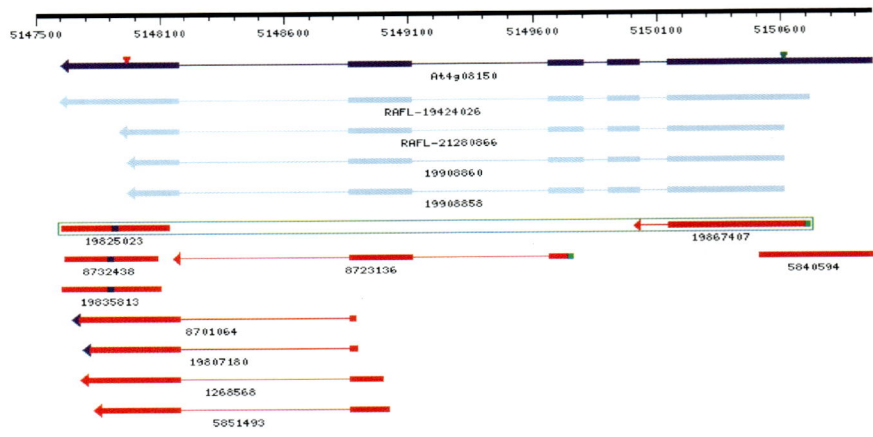

**Predicted gene structure (within gDNA segment 5148150 to 5149760):**

Ex•n  1    5149760 5149660 (101 n); EST  1 101 ( 101 n); score:  1.000
In•ron 1  5149659 5149113 (547 n); Pd:  0.994 (s: 1.00), Pa: 0.999 (s: 1.00)
Ex•n  2    5149112 5148856 (257 n); EST 102 358 ( 257 n); score:  1.000
In•ron 2  5148855 5148173 (683 n); Pd:  0.990 (s: 1.00), Pa: 0.974 (s: 0.00)
Ex•n  3    5148172 5148150 ( 23 n); EST 359 381 ( 23 n); score:  1.000

PG5_ATCHR4-_8723136+    (5149760 5149660,5149112 5148856,5148172 5148150)

**Alignment:**
```
GCTAACAAGG CCCATTCAGG AAGCAATGGA GTTTATACGT CGTATTGAAT CTCAGCTTAG 5149701
|||||||||| |||||||||| |||||||||| |||||||||| |||||||||| ||||||||||
GCTAACAAGG CCCATTCAGG AAGCAATGGA GTTTATACGT CGTATTGAAT CTCAGCTTAG   60

CATGTTGTGT CAGAGTCCCA TTCACATCCT CAACAATCCT GGTACATGTC ATCAATCTCA 5149641
|||||||||| |||||||||| |||||||||| |||||||||| |
CATGTTGTGT CAGAGTCCCA TTCACATCCT CAACAATCCT G.......... ..........  101

///

TAAATGACTA AACCATTTAA CTTCATGAAA TATATTATGT TGAAGCAGAT GGGAAGAGTG 5149101
                                                    || ||||||||||
.......... .......... .......... .......... ........AT GGGAAGAGTG  113

///

CTTGTGTACG TATAAATTTT CTCTCATCTC TCTTTGCCCT CATATTTTTT AAAATATATA 5148801
|||||
CTTGT..... .......... .......... .......... .......... ..........  358

///

AAGATGGTTG GTTTGAGTTG TGTGGCAGGA GTCAGAGAAG GTAGCGTTGG C           5148150
                                  || |||||||||| |||||||||| |
.......... .......... ........GA GTCAGAGAAG GTAGCGTTGG C            381
```

**FIGURE 1. A.** Schematic representation of spliced alignment evidence for a gene structure annotation in the *Arabidopsis thaliana* genome. The display was generated at the AtGDB web site (http://www.plantgdb.org/AtGDB/), where analogous displays can be viewed for the entire genome. The scale (top bar) and the dark blue gene structure represent the current chromosome assembly and gene annotation for chromosome four (GenBank accession NC_003075; Wortman et al., 2003). Solid boxes correspond to exons and thin lines correspond to introns. The 5′- and 3′-boundaries of the coding region are indicated by green and red triangles, respectively. Arrows indicate the direction of transcription, which in case for multi-exon spliced alignments is inferred from the implied splice site patterns (Usuka et al., 2000). Full-length cDNA spliced alignments are shown in light blue. EST spliced

any given sequence segment have not been successful to produce entirely reliable gene structures (reviewed by Pavy et al., 1999). A complementary approach is to use cDNA or EST evidence in "spliced alignments" that delineate exons and introns (Brendel and Zhu, 2002). A number of programs are available to derive such alignments, including BLAST-like methods such as sim4 (Florea et al., 1998), Spidey (Wheelan et al., 2001), or BLAT (Kent, 2002) and dynamic programming methods such as dds/gap2 (Huang et al., 1997) or GeneSeqer (Usuka et al., 2000). Zhu et al. (2003) reported a large-scale application of the GeneSeqer program to align ESTs to the *Arabidopsis* genome, which revealed many cases of wrong gene structure annotation and provided additional information on untranslated regions and instances of alternative splicing. The latest version of the *Arabidopsis* genome annotation also incorporates comprehensive cDNA/EST evidence, largely based on an automated pipeline for spliced alignment (Wortman et al., 2003; Haas et al., 2003). Limitations of this approach mostly result from insufficient cDNA/EST sampling to annotate entire gene structures and all genes and technical problems for at least some of the programs to correctly identify short exons and introns with non-canonical splice sites (Haas et al., 2002).

Figure 1 gives an example of spliced alignment display at the AtGDB web site (http://www.plantgdb.org/AtGDB/). This site was set up to allow viewing of all available cDNA/EST evidence for *Arabidopsis* gene structure annotations (Zhu et al., 2003). For this example, the evidence supports the current annotation, and thus this gene model can be confidently accepted and should be classified as a fully confirmed model. In the next section I will discuss examples from a currently large list of gene structure models that are contradicted by the cDNA/EST evidence. Because sophisticated annotation pipelines are already in use to automatically incorporate such evidence into the annotation (Haas et al., 2003), it would seem that alternatives to automated annotation must be contemplated. I will discuss tools for user contributed annotations available at AtGDB and argue that such approaches ought to be tried as attempts to involve the expert community in improving the accuracy of genome annotation for the benefit of the entire community.

←——————————————————————————————————————

**FIGURE 1. (*continued*)** alignments are in red. Multi-exon 5′-ESTs are marked by green color at their 5′-terminus, and multi-exon 3′-ESTs are marked by blue color at their 3′-terminus. Single exon ESTs have corresponding 5′ / 3′ labels at the center of their representations. Pairs of 5′- and 3′-ESTs from the same clone are grouped by green boxes. Numbers labeling the spliced alignments are GenBank gi tags for the corresponding sequences. The RAFL tag identifies full-length cDNAs derived from the Seki et al. (2002) project. The gene (At4g08150) encodes a knotted-1 like homeobox protein. EST evidence supports all but the second intron. cDNA evidence supports the entire gene structure annotation, although the 5′- and 3′-termini of the transcripts are either variable or not always captured in the cDNAs. **B.** Partial GeneSeqer output of the spliced alignment of EST gi:8723136 (plus strand) to the genomic region on chromosome four displayed in **A**. The spliced alignment entails three exons as summarized on top. Exon scores are normalized sequence similarity scores. Pd, donor site score; Pa, acceptor site score (the s-values in parenthesis are the normalized sequence similarity scores in the adjacent 50 exon nucleotides). Only part of the alignment is shown (omitted parts indicated by ///). Identities between the genomic sequence (upper lines) and the EST sequence (lower lines) are indicated by vertical bars. Introns are represented by dots. All alignments are viewable at the AtGDB web site.

## 5. THE CASE FOR A COMMUNITY APPROACH TO ANNOTATION

The annotation of At4g08150 displayed in Figure 1 represents a nice case from an annotator's perspective: the cDNA/EST evidence is consistent with the *ab initio* predicted gene structure and simply extends that gene structure to include the 5'- and 3'-untranslated regions of the mRNA transcripts. Future evidence may of course indicate additional transcription initiation and termination sites or alternative splice sites, however the current annotation will stand as representing true transcripts.

Unfortunately, or rather as a matter of fact, there are many cases for which the annotation problem is more challenging. A first example is displayed in Figure 2. The AtGDB display of spliced alignments shows a contradiction between the first annotated intron of the At1g24050 gene and the intron borders confirmed by spliced alignment of two cDNAs and five ESTs in the same region. Inspection of the spliced alignments (not shown) reveals the presence of an AT-AC U12-type intron (Zhu and Brendel, 2003). This suggests that the annotated gene structure was based on an *ab initio* prediction using standard programs that are not able to predict introns other than standard GT-AG U2-type introns, and thus the annotation includes the best fitting GT-AG intron to extend the open reading frame across the first two exons. Because there is no evidence for the *ab initio* predicted U2-type intron, the At1g24050 gene structure annotation should be corrected.

How should annotation corrections be implemented? Because the *Arabidopsis* genome project, including annotation and database management, probably is as well funded as can be expected and involves state-of-the-art bioinformatics support, it stands to reason that automated, computational annotation will not be adequate to resolve more challenging annotation problems.

Neither should we expect that a small group of staff scientists will be able to look at all gene models individually (numbering approximately 27,000–30,000 for *Arabidopsis*!). Thus, my group has been exploring pathways to community-based annotation, with a focus on exploring web-based technologies to enable such effort. To illustrate, I have used the AtGDB User Contributed Annotation tools to communicate the corrected At1g24050 gene structure. Using the "Login" function at AtGDB (upper left corner, Figure 2), the contributor will see a screen similar to the one displayed in Figure 3. My contributed annotation as entered via the web-form is shown in Figure 4. The schematic gene structure is displayed in green. Links to the genomic context and the newly created database record are provided above the gene structure, and details of the annotation have been entered into text boxes below. Once saved, the user-contributed annotation is displayed along with the current GenBank annotation and all the cDNA/EST evidence as before (Figure 5). Until AtGDB staff approves this annotation, it will only be visible to the user who contributed it. Once approved, it will be visible to anyone over the web. Periodically, such annotations with their associated evidence could be downloaded by other *Arabidopsis* databases and incorporated into

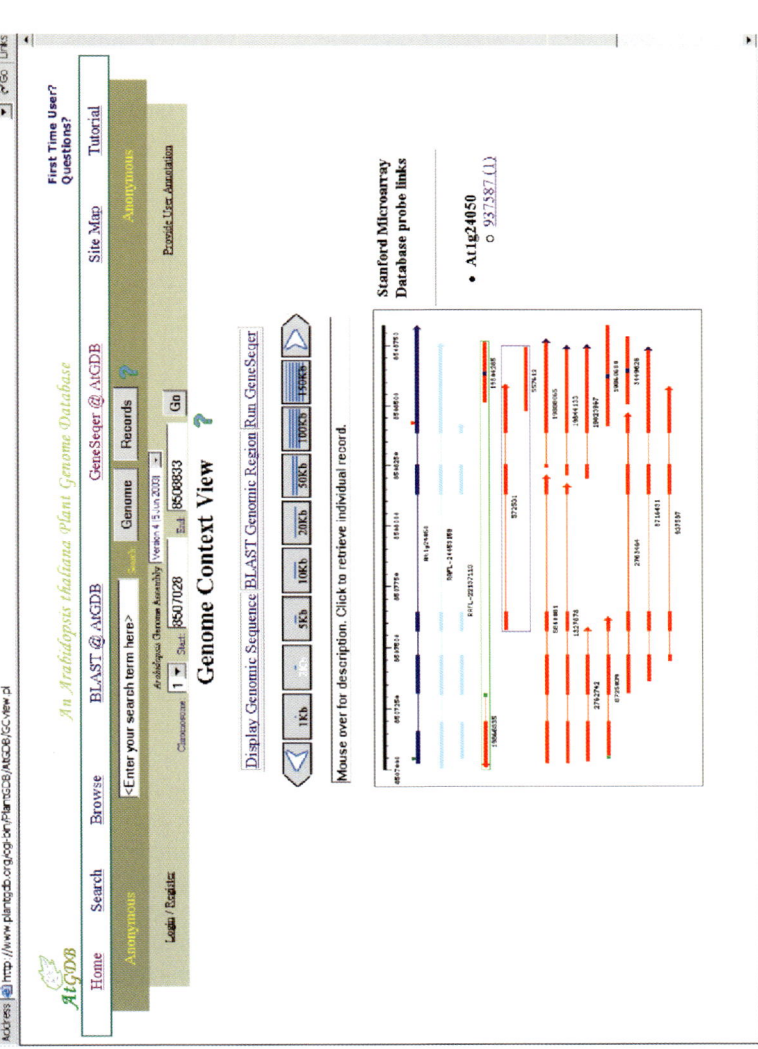

**FIGURE 2.** AtGDB display of the *Arabidopsis thaliana* chromosome one region containing the At1g24050 gene model. The top toolbar indicates the extent of the region being displayed and the version of annotation. Note the user name "Anonymous" and the Login/Register/Provide User Annotation functions below. The spliced alignment display (color scheme and labels as in Figure 1) indicates a discrepancy between the GenBank annotation and the cDNA/EST evidence for intron one. Inspection of the spliced alignments (not shown) reveals a U12-type intron in place of the annotated U2-type intron. Because there is no cDNA/EST evidence that supports the annotated intron, it presumably was derived from *ab initio* computational predictions only.

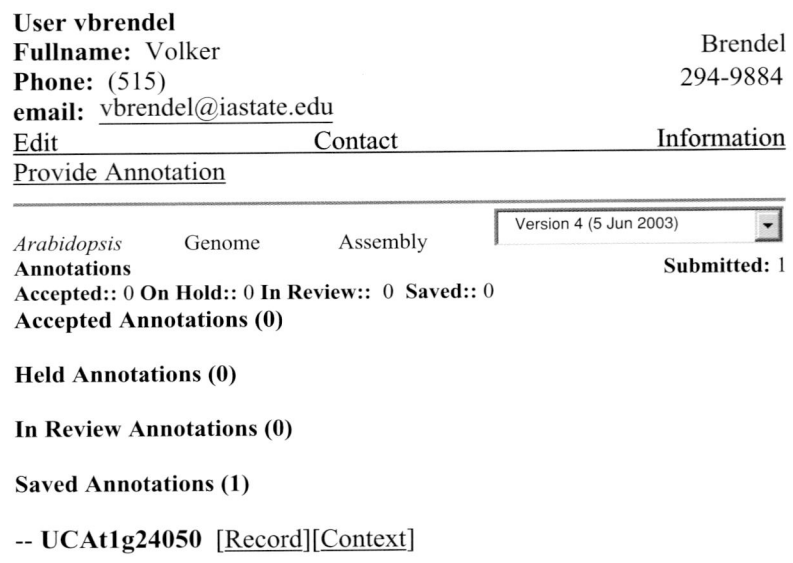

**User vbrendel**
**Fullname:** Volker                                                          Brendel
**Phone:** (515)                                                          294-9884
**email:** vbrendel@iastate.edu
Edit                                    Contact                                    Information
Provide Annotation

*Arabidopsis*      Genome      Assembly      | Version 4 (5 Jun 2003) |  ▾
**Annotations**                                                          **Submitted:** 1
**Accepted::** 0 **On Hold::** 0 **In Review::**  0  **Saved::** 0
**Accepted Annotations (0)**

**Held Annotations (0)**

**In Review Annotations (0)**

**Saved Annotations (1)**

**-- UCAt1g24050**  [Record][Context]

**FIGURE 3.** Sample user profile for maintaining user contributed gene structure annotations at AtGDB. This screen is reached after login with a registered user name, using the login functions at any AtGDB screen. Users can review their contributed annotations, which are classified according to the level of curation (accepted annotations have been reviewed by AtGDB staff and are visible to any web user of AtGDB).

future versions of the public genome annotation. The above illustration would be mostly an exercise if it were not for the fact that a large number of the *Arabidopsis* gene structure annotations remain incorrect, a problem that will be compounded for less mature sequencing projects like rice or maize. Figures 6 and 7 illustrate other typical problems. Full-length cDNA spliced alignment indicates that the two gene models At2g40835 and At2g40840 actually correspond to a single gene (Figure 6). In this case, the fifth intron of the cDNA spliced alignment is an AT-AC U12-type intron, and here *ab initio* prediction simply inserted an intergenic region in its place. Figure 7 shows a frequent case of wrongly annotated overlapping 3'-ends of genes on opposite strands. This annotation problem is caused by automated assembly of EST spliced alignments into gene models and the inability of this procedure to associate single-exon ESTs in the region of potential overlap to the correct gene model (Haas et al., 2003). In this case, full-length cDNA evidence clearly indicates the correct 3'-ends, but one should appreciate the difficulty in writing procedures that would automatically give correct answers in all such cases and their variants. The full-length cDNA spliced alignment spans the gene models At2g40835 and At2g40840. These two annotations should be replaced by a single gene structure (encoding a disproportionating enzyme in starch metabolism).

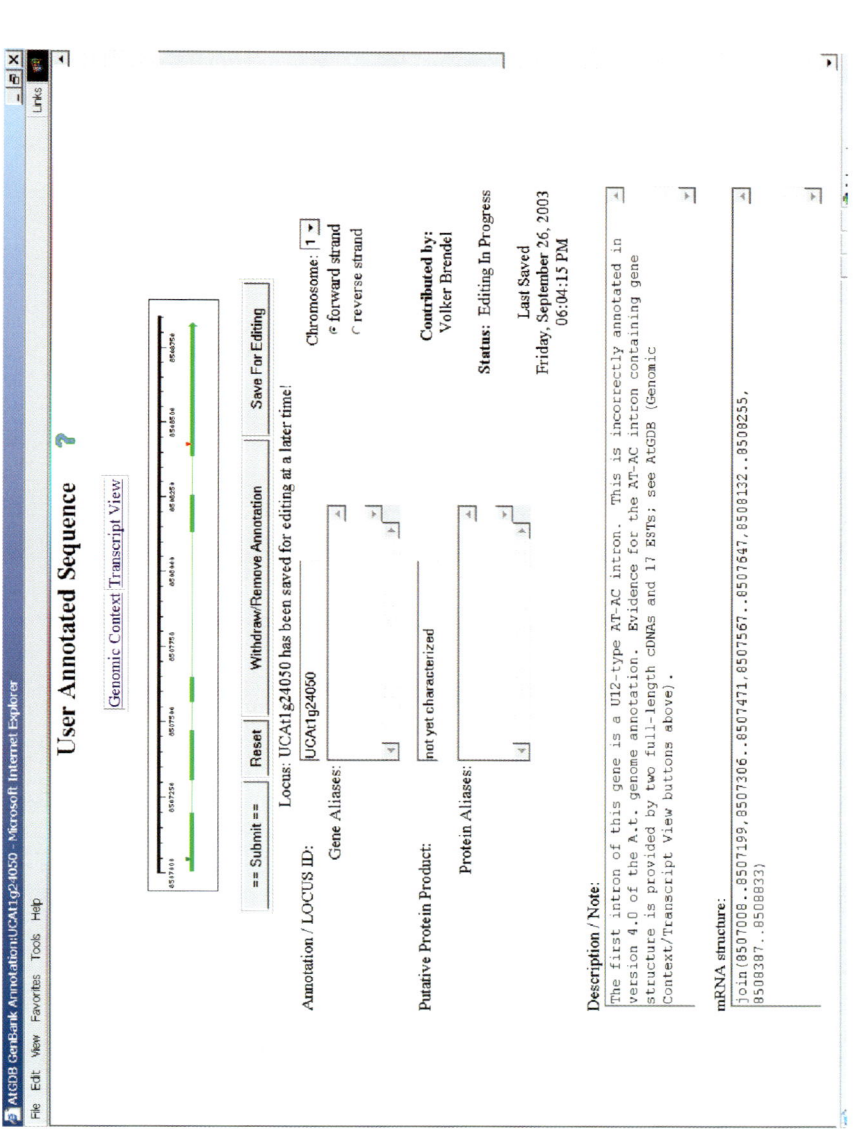

**FIGURE 4.** Sample gene structure annotation screen at AtGDB. Based on the cDNA/EST evidence displayed in Figure 2, a corrected gene structure was entered in GenBank style annotation and documented.

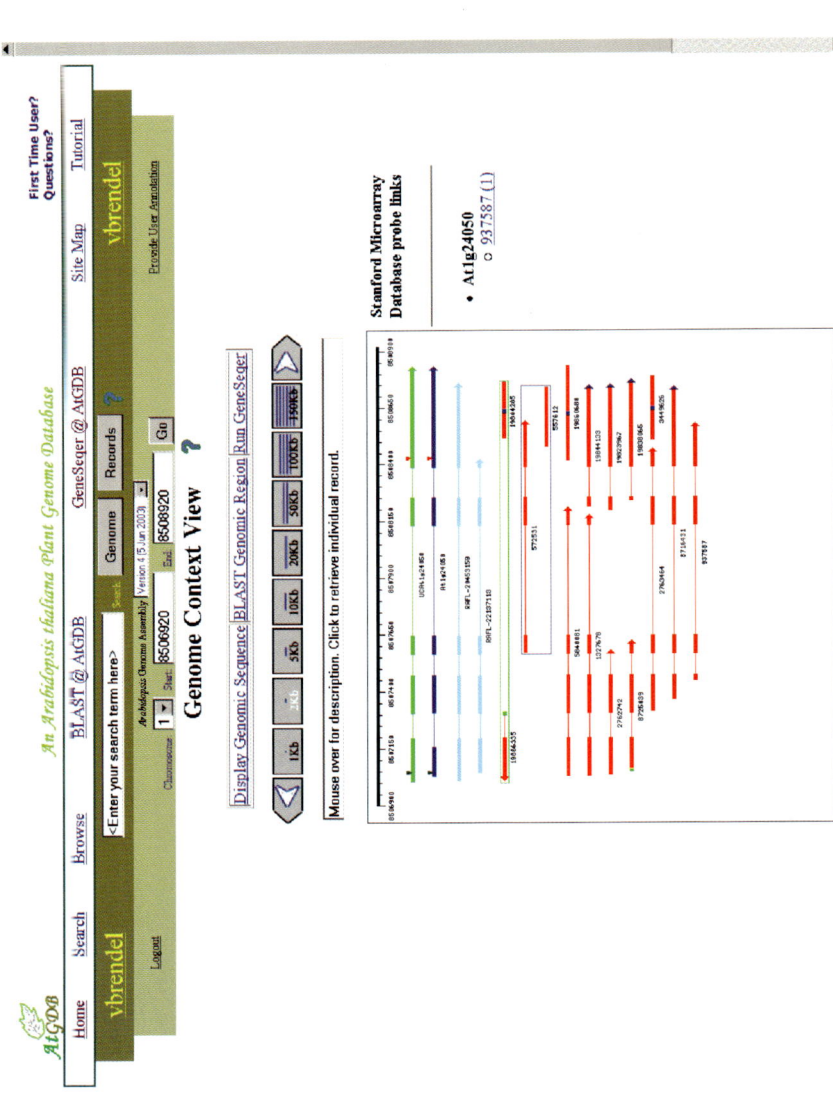

**FIGURE 5.** AtGDB display of the At1g24050 region after user contributed annotation. Compared to Figure 2, note the changed user name in the toolbar and the additional gene structure display labeled UCAt1g24050, which is consistent with the cDNA/EST evidence and should replace the current GenBank annotation.

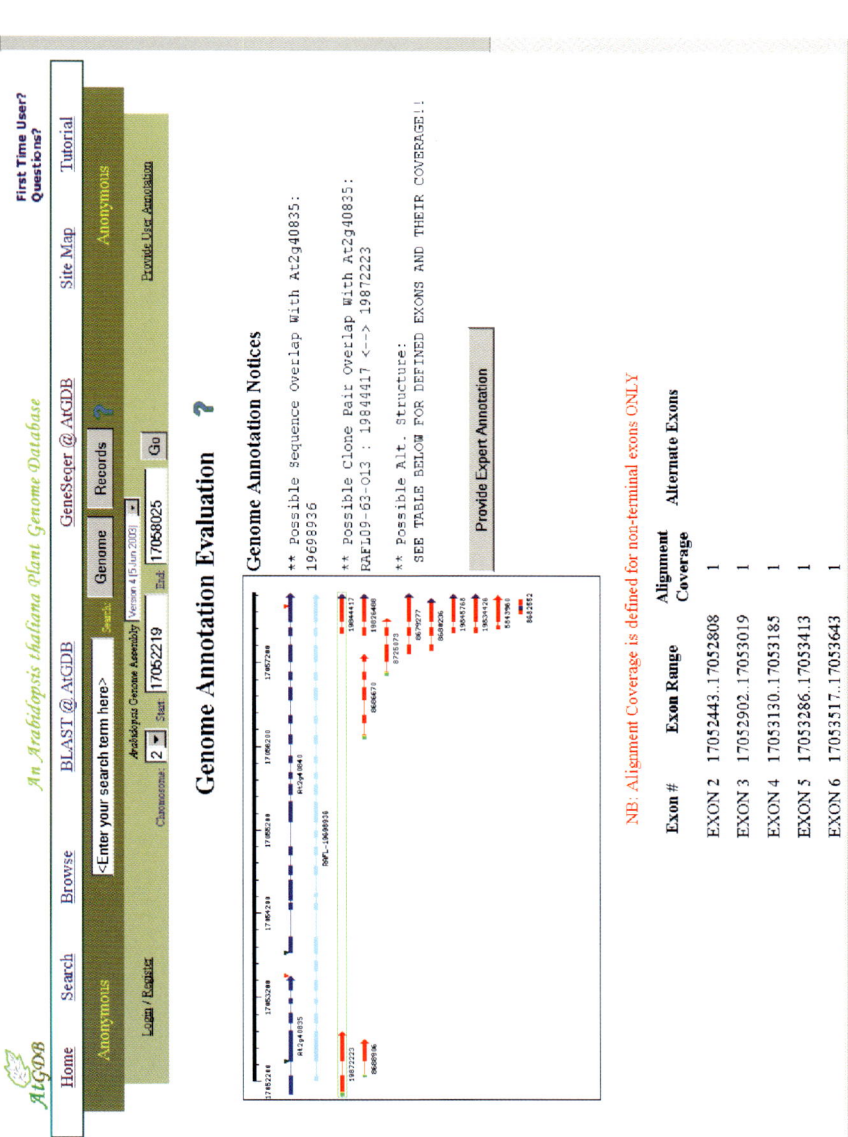

**FIGURE 6.** AtGDB display of a typical annotation error involving incorrect assignment of an intergenic region.

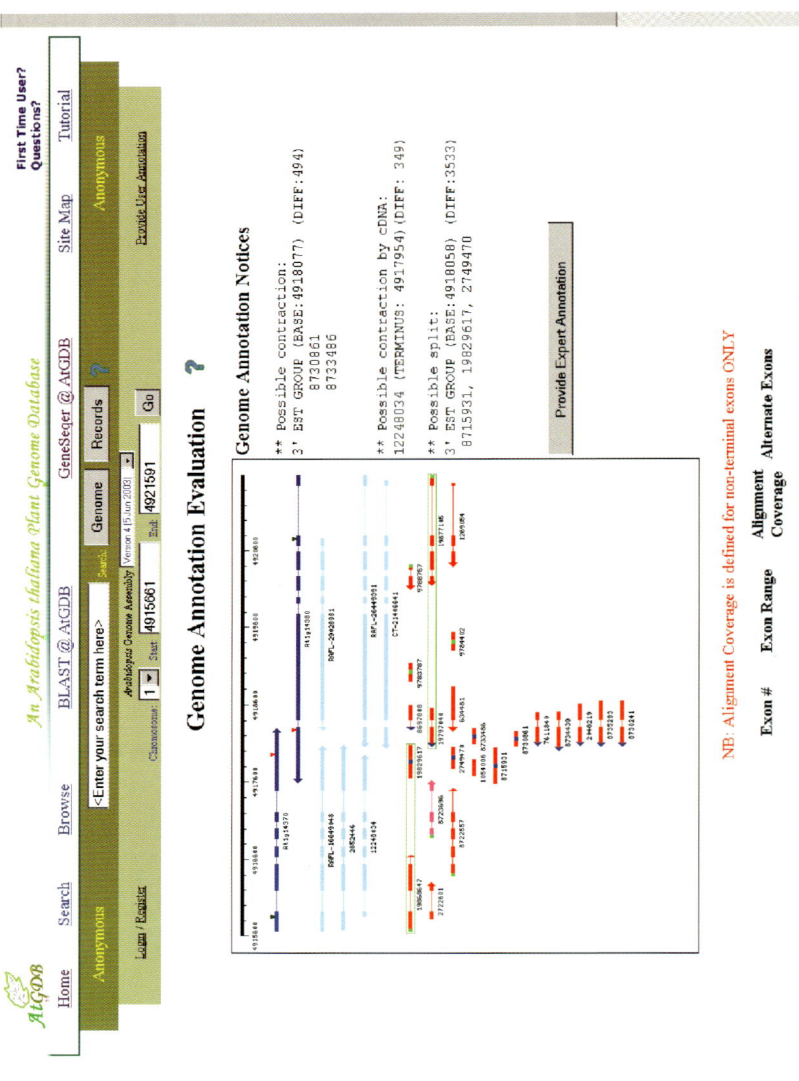

**FIGURE 7.** AtGDB display of a typical annotation error involving incorrect assignment of 3′-untranslated regions. Spliced alignment of full-length cDNAs clearly separates the At1g14370 and At1g14380 genes. Automated annotation as currently implemented (Haas et al., 2003) incorrectly assigns some of the 3′-terminal single-exon ESTs from At1g14370 also to the At1g14380 model on the opposite strand.

74                                        **Volker Brendel**

## 6. PERSPECTIVE

I hope I have put forward convincing arguments for community-base genome annotation efforts. It seems to me that there are three factors that necessitate serious exploration of community-based efforts. These factors are, first, the inherent difficulties with automated, computational gene structure annotation (at least at this stage of our theoretical understanding of gene structure); second, the impracticality of reliance on a small staff of dedicated curators to encompass all annotation needs of a large eukaryotic genome; and, third, the need for accurate gene structure annotation as a foundation for all other functional and evolutionary studies. Of course, current annotation providers are already open to and in fact soliciting user comments. However, for large-scale and systematic efforts to be successful, a computational infrastructure must be put in place that will make user-contributions easy, as standardized as possible, and inclusive of evidence that can quickly be verified by curatorial database staff. We have recently shown that the GeneSeqer spliced alignment program can successfully employ ESTs from multiple related species to predict gene structure (Schlueter et al., 2003). This suggests that this approach is promising for the annotation of more complex plant genomes such as rice, maize, and legume species that will be sequenced in the next few years.

ACKNOWLEDGMENTS. This work was supported in part by NSF Plant Genome Research Projects grants DBI-0110254 and DBI-9872657. I am indebted to my research group at Iowa State University for their excellent work and friendship. The *Arabidopsis* web-based gene structure annotation system has been pioneered by graduate student Shannon D. Schlueter.

## 7. REFERENCES

The Arabidopsis Genome Initiative, 2000, Analysis of the genome sequence of the flowering plant *Arabidopsis thaliana*, *Nature* **408**:796–815.

Brendel, V., 2002, Integration of data management and analysis for genome research. In S. Schubert, B. Reusch and N. Jesse (eds.), "Informatik bewegt". *Lecture Notes in Informatics (LNI) – Proceedings* **P-20**:10–21.

Brendel, V., and Zhu, W., 2002, Computational modeling of gene structure in *Arabidopsis thaliana*, *Plant Mol. Biol.* **48**:49–58.

Campbell, A.M., and Karlin, S., 2003, Evolution of genome structures, *Theor. Pop. Biol.* **61**:365–366.

Florea, L., Hartzell, G., Zhang, Z., Rubin, G.M., and Miller, W., 1998, A computer program for aligning a cDNA sequence with a genomic DNA sequence, *Genome Res.* **8**:967–974.

Haas, B., Delcher, A.L., Mount, S.M., Wortman, J.R., Smith, R.K. Jr., Hannick, L.I., Maiti, R., Ronning, C.M., Rusch, D.B., Town, C.D., Salzberg, S.L., and White, O., 2003, Improving the *Arabidopsis* genome annotation using maximal transcript alignment assemblies, *Nucl. Acids Res.* **31**:5654–5666.

Haas, B.J., Volfovsky, N., Town, C.D., Troukhan, M., Alexandrov, N., Feldman, K.A., Flavell, R.B., White, O., and Salzberg, S.L., 2002, Full-length messenger RNA sequences greatly improve genome annotation.*Genome Biol.* **3**:research0029.1-0029.12.

Huang, X., Adams, M.D., Zhou, H., and Kerlavage, A.R., 1997, A tool for analyzing and annotating genomic sequences, *Genomics* **46**:37–45.

Kent, W.J., 2002, BLAT–the BLAST-like alignment tool, *Genome Res.* **12**:656–664.

Konopka, A., 2003, Selected dreams and nightmares about computational biology, *Comp. Biol. & Chem.* **27**:91–92.

Korning, P.G., Hebsgaard, S.M., Rouzé, P., and Brunak, S., 1996, Cleaning the GenBank *Arabidopsis thaliana* data set, *Nucl. Acids Res.* **24**:316–320.

The National Plant Genomics Initiative: objectives for 2003–2008, *Plant Physiol.* **130**:1741–1744.

Pavy, N., Rombauts, S., Déhais, P., Mathé, C., Ramana, D.V., Leroy, P., and Rouzé, P., 1999, Evaluation of gene prediction software using a genomic data set: application to *Arabidopsis thaliana* sequences, *Bioinformatics* **15**:887–899.

Schlueter, S.D., Dong, Q., and Brendel, V., 2003, GeneSeqer@PlantGDB—gene structure prediction in plant genomes, *Nucl. Acids Res.* **31**:3597–3600.

Seki, M., Narusaka, M., Kamiya, A., Ishida, J., Satou, M., Sakurai, T., Nakajima, M., Enju, A., Akiyama, K., Oono, Y., et al., 2002, Functional annotation of a full-length *Arabidopsis* cDNA collection, Science **296**:141–145.

Usuka, J., Zhu, W., and Brendel, V., 2000, Optimal spliced alignment of homologous cDNA to a genomic DNA template, *Bioinformatics* **16**:203–211.

Wheelan, S.J., Church, D.M., and Ostell, J.M., 2001, Spidey: a tool for mRNA-to-genomic alignments. *Genome Res.* **11**:1952–1957.

Wortman, J., Haas, B.J., Hannick, L.I., Smith, R.K. Jr., Maiti, R., Ronning, C.M., Chan, A.P., Yu, C., Ayele, M., Whitelaw, C.A., White, O.R., and Town, C.D., 2003, Annotation of the Arabidopsis genome, *Plant Physiol.* **132**:461–468.

Zhu, W., and Brendel, V., 2003, Identification, characterization, and molecular phylogeny of U12-dependent introns in the *Arabidopsis thaliana* genome, *Nucl. Acids Res.* **31**:4561–4572.

Zhu, W., Schlueter, S.D., and Brendel, V., 2003, Refined annotation of the *Arabidopsis thaliana* genome by complete EST mapping, *Plant Physiol.* **132**:469–484.

*Chapter 6*

# FCModeler

## Dynamic Graph Display and Fuzzy Modeling of Metabolic Maps

Julie A. Dickerson

## 1. MODELING METABOLIC NETWORKS

Metabolic networks combine metabolism and regulation. These complex networks are difficult to understand and visualize due to the diverse types of information that need to be represented. FCModeler, a publicly available software package is designed to enable the biologist to visualize and model metabolic and regulatory network maps in plants. It links to an interactions database (MetNetDB) containing information on regulatory and metabolic interactions derived from a combination of web databases and input from biologists in their area of expertise. FCModeler displays input from MetNetDB in a graphical form. Sub-networks can be identified and interpreted using fuzzy cognitive maps. FCModeler is intended to develop and evaluate hypotheses, and provide a modeling framework for assessing the large amounts of data captured by high-throughput gene expression experiments. Three-dimensional graph visualization coupled with visualization of the physical structure of a cell helps create a novel integrated information workspace for the study of metabolic networks.

A major challenge in the post-genome era is to understand how interactions among molecules in a cell determine its form and function. With the help of

**Julie A. Dickerson**    Department of Electrical and Computer Engineering, Iowa State University, Ames, Iowa 50010

*Genome Exploitation: Data Mining the Genome*, edited by J. Perry Gustafson, Randy Shoemaker, and John W. Snape.
Springer Science + Business Media, New York, 2005.

transcriptomic, proteomic and metabolomic analysis technologies, biologists can obtain vast amounts of valuable data on metabolic network interactions, and many approaches are being developed to analyze the resultant data (Brown et al., 2000; Dickerson et al., 2002; Dougherty et al., 2002; Oliver et al., 2002; Yao 2002). Several of these approaches use complex databases of cellular interactions. The WIT Project (Overbeek et al. 2000) (http://wit.mcs.anl.gov/WIT2/WIT) and the Kyoto Encyclopedia of Genes and Genomes (Kanehisa et al. 2000) (KEGG) (http://www.genome.ad.jp/kegg) provide molecular networks based on prokaryotic metabolism. WIT produces reconstructions of the metabolism of the organism derived from sequence, biochemical, and phenotypic data, organized as a static presentation. KEGG's goals are to computerize existing knowledge of the information pathways that consist of interacting genes or molecules and to link individual components of the pathways with the gene catalogs being produced by the genome projects. This approach provides the framework for eventual simulations. EcoCyc is a pathway/genome database for *Escherichia coli* that describes its enzymes, and its transport proteins. MetaCyc describes pathways and enzymes for many different organisms (Karp et al., 2002; Karp et al., 2002). The databases combine information from a number of sources and provide function-based retrieval of DNA or protein sequences.

MetNetDB combines information from AraCyc, a set of Arabidopsis specific biochemical pathways derived from the MetaCyc model and curated by an expert in Arabidopsis (http://aracyc.stanford.edu/). MetNet is designed to provide a framework for the formulation of testable hypotheses regarding the function of specific genes, proteins, and metabolites, and in the long term provide the basis for identification of genetic regulatory networks that control plant composition and development (Dickerson et al., 2001; Dickerson et al., 2002; Ding et al., 2002; Oliver et al., 2002). Our primary focus is on the eukaryotic model plant, Arabidopsis (http://www.arabidopsis.org/). The entire Arabidopsis genome has been sequenced, databases cataloging genes and gene function are expanding, and new databases, such as protein-protein interactions, are being initiated. However, more than half of the 26,000 Arabidopsis genes have no assigned function, and many of the remaining genes have only putatively assigned biochemical functions. Even less is understood about the metabolic, structural or regulatory role of each gene product, its interactions with other cellular components, and the kinetics of each interaction. Our approach to reveal complex biological networks is to extract information from gene expression data sets and combine it with what is known about metabolic and regulatory pathways to achieve a better understanding of how metabolism is regulated in a eukaryotic cell.

## 2. METABOLIC NETWORKING DATABASE

The Metabolic Networking DataBase (MetNetDB) contains a metabolic and regulatory map of Arabidopsis with a user-friendly JAVA interface for creating and searching the map. The map, together with gene expression data (metabolomics, proteomics, and microarray), can be transferred to FCModeler as an XML file, for use in data exploration.

The MetNetDB map is being assembled by biologists with expertise in specific areas of metabolism. It is composed of entities (genes, RNAs, polypeptides, protein complexes, metabolites, and environmental inputs) connected by interactions (conversion, catalytic, regulatory). Identities of the genes, RNAs, and polypeptides have been downloaded from TAIR (http://www.arabidopsis.org/). Protein complexes are currently added by expert users, as there is no adequate database of protein complexes in Arabidopsis. Identities of many metabolites have been downloaded from KEGG (http://www.genome.ad.jp/kegg/); metabolites not present in KEGG are manually added as new entities by expert users, based on their CAS registry number (http://www.cas.org/EO/regsys.html).

The metabolic reactions from the AraCyc database have been downloaded into MetNetDB. An important aspect of the map is the inclusion of information on subcellular location. This is critical, because particular entities can interact contingent on being located in the same subcellular compartment. A given entity may be present as separate pools in multiple compartments, for example citrate is present in the mitochondria (where it participates in the TCA cycle) and the cytosol (where it is a substrate for cytosolic acetyl-CoA formation (Fatland et al. 2002)).

## 3. MODELING METABOLIC RELATIONSHIPS

Three basic types of interactions are conversion, regulatory, and catalytic. In a conversion interaction, a node (typically a chemical(s)) is converted into another node, and used up in the process. A catalytic interaction represents an enzyme that enables a chemical conversion and does not get used up in the process. In a regulatory interaction, the entity activates or deactivates another node, and is not used up in the process.

A wide variety of cellular processes can be represented, each occurring to entities in specified subcellular compartments. For example, to represent the reaction catalyzed by ATP citrate lyase (ACL), that generates cytosolic acetyl-CoA (Fatland et al. 2002), two interactions are used. One is a conversion interaction; its inputs are $citrate_{cytosol} + CoA_{cytosol} + ATP_{cytosol}$ and its outputs are $acetyl\text{-}CoA_{cytosol} + oxaloacetic\ acid_{cytosol} + ADP_{cytosol} + PO_{4cytosol}$. The second is a catalytic interaction; its input is ATP citrate $lyase_{cytosol}$. In another example, to represent the translocation of citrate from the mitochondrion to the cytosol, two entities and a single conversion interaction are used: $citrate_{mitochondrion}$ goes to $citrate_{cytosol}$. The formation or modification of a protein complex can be represented. For example, ACLA and ACLB are the subunits that compose the enzyme ACL. A single conversion interaction is used to represent the reaction; its inputs are $ACLA_{cytosol}$, and $ACLB_{cytosol}$. Its output is $ACL_{cytosol}$.

### 3.1. Graphing the Metabolic Map: FCModeler

The main goals of the FCModeler package are to capture the intuitions of biologists and provide a modeling framework for assessing large amounts of information and to test the effects of hypotheses. The tools that are being developed use graph theoretic approaches to analyze network structure and behavior and fuzzy methods that model changes in the network (Dickerson et al., 2002). There are three parts of this system: a dynamic graph visualization package written in Java, graph-theoretic analysis to find critical paths, and modeling using fuzzy cognitive maps to capture uncertainty in the model. Figure 1 shows a sample sub-graph from the MetNetDB for OAA metabolism in Arabidopsis and highlights some of the visualization flexibility available in FCModeler.

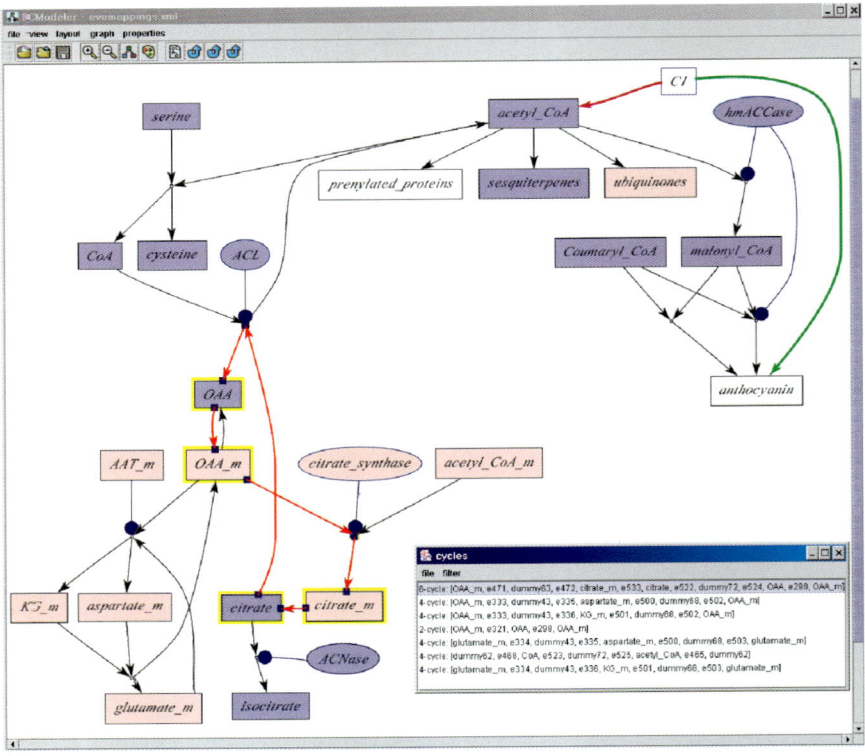

**FIGURE 1.** The highlighted nodes and links show a small cycle within the metabolic network. The text box gives a list of all cycles found in the displayed graph. Colors and shapes of features can be user-designated: entities in mitochondria and cytosol are shaded; entities in unknown location, white; enzymes are shaped as ellipses.

## 3.2. Modeling Metabolic Networks Using Fuzzy Cognitive Maps

The interactions (also referred to as edges or links) in the network are modeled as fuzzy functions depending on the detail known about the network. Modeling using fuzzy cognitive maps (FCMs) is performed in the Matlab™ analysis program and the results showing node activation levels are animated in FCModeler. Fuzzy cognitive maps are fuzzy digraphs that model causal flow between concepts (Dickerson et al., 1994) or, in this case, biomolecular entities (Dickerson et al., 2001; Dickerson et al., 2002). Entities stand for causal fuzzy sets where events occur to some degree. The entities are linked by interactions that show the degree to which these entities depend on each other. Interactions stand for causal flow. The sign of an interaction (+ or −) shows causal increase or decrease between entities. The fuzzy structure allows the RNA, metabolite, or protein levels to be expressed as continuous values. This modeling has demonstrated regulation in the Arabidopsis network, in the case of gibberellin conversion from an inactive form to an active form (Dickerson et al., 2001).

Fuzzy cognitive maps (FCMs) have the potential to answer many of the concerns that arise from the existing models. Fuzzy logic allows a concept or gene expression to occur to a degree—it does not have to be either on or off (Kosko 1986). FCMs have been successfully applied to systems that have uncertain and incomplete models that cannot be expressed compactly or conveniently in equations. Some examples are modeling human psychology (Hagiwara, 1992), and on-line fault diagnosis at power plants (Lee et al., 1996). All of these problems have some common features. The first is the lack of quantitative information on how different variables interact. The second is that the direction of causality is at least partly known and can be articulated by a domain expert. The third is that they link concepts from different domains together using arrows of causality. These features are shared by the problem of modeling the signal transduction and gene regulatory networks.

Simple or trivalent FCMs have causal edge weights in the set $\{-1, 0, 1\}$ and concept values in $\{0, 1\}$ or $\{-1, 1\}$. Simple FCMs give a quick approximation to an expert's causal knowledge. More detailed graphs can replace this link with a time-dependent and/or nonlinear function. The types of link models used in the current project are described below.

*Regulatory Links:* The regulatory edges are modeled using a simple FCM model that assumes binary connecting edges for the single edge case. When there are multiple excitatory or inhibitory connections, the weights are divided by the number of input connections in the absence of other information. As more information becomes known about details of the regulation, for example how RNA level affects the translation of the corresponding protein, the function of the link models will be updated. The regulatory nodes will also have self-feedback since the nodes stay on until they have been inhibited.

*Conversion Links:* Conversion relationships are modeled in different ways depending on the goal of the simulation study. The first case corresponds to investigating causal relationships between nodes. The node is modeled in the same manner as a regulatory link in which the presence of one node causes presence at the next node. When information about the rate of change in a reaction is available, a simple difference equation can model the gradually rising and falling levels of the nodes.

*Catalyzed Links:* Catalyzed reactions add a dummy node that acts upon a conversion link. This allows one link to modify another link. In the current model, the catalyzed link is simulated by weighting the input node in such a way that both inputs must be present for the node to be active. Another method of modeling catalyzed links is an augmented matrix that operates on the edges between the nodes. The catalyst node acts as a switch that allows a reaction to occur in the proper substrates are available.

### 3.3. Cycle Analysis in Metabolic Networks

Complex metabolic networks can be analyzed by searching for paths between two entities and assessing the effects that the entities have on each other or by searching for feedback cycles in the network as shown in Figure 1. Graph theoretic methods are a promising path for characterizing metabolic networks and discover subsystems for detailed modeling. Graph analysis such as searching for alternate paths and feedback cycles can give information about the underlying biology. Preliminary results have found existing pathways in the map as well as new relationships in the models.

Cycles in the data show repeated patterns in the network. These cycles range from simple loops in which a gene causes a protein to be expressed, when the protein is present to a degree, the gene is turned off or down. More complex cycles encompass entire metabolic pathways in an organism. The interactions or overlaps between the cycles show how these control paths interact. There are three basic types of cycle in metabolic and regulatory pathways:

Gene expression cycle—a protein A is a part of a pathway that controls expression of a gene B, gene B produces protein B, then protein B cycles back into the pathway which contains protein A, at a point upstream of protein A.

Signal transduction cycle—contains no genes, e.g., receptor A activates a downstream sequence of signal transduction such as: A$\rightarrow$ B$\rightarrow$ C$\rightarrow$ D$\rightarrow$ A where D feeds back to inhibit A. These types of cycles may play a key role in considering pathways as targets for drugs.

Metabolic feedback cycle—a small molecule from a metabolic pathway diffuses upstream to activate or inhibit an enzyme involved in its production.

## 4. REPRESENTING METABOLIC PATHWAYS IN VIRTUAL REALITY

Complex interactive metabolic pathways contain many different types of information, which presents a challenge to computationally model and visualize the interactions. Two-dimensional graph-based models of metabolic networks are overloaded since edges and nodes can have multiple meanings (Dickerson et al., 2002). Three-dimensional graph visualization coupled with visualization of the physical structure of a cell help create a novel integrated information workspace for the study of metabolic networks.

Virtual reality (VR) and immersive environments are relatively new research tools. VR strives to present the user with a convincing, interactive three-dimensional (3-D) environment. The user views the 3-D environment stereoscopically, typically with the aid of specialized glasses. The user's position is tracked by a computer so the virtual environment can respond to the user's movements. Projection-based virtual reality systems, such as Iowa State University's enclosed cube structure called the C6, have stereo images projected onto the display surfaces (Cruz-Neira et al., 1993).

Creating a virtual metabolic network environment requires visualizing and navigating complex graphs in a scalable immersive environment, and integrating physical models and graphical representations of cell metabolism. The goal is to create a seamless system to enable biologists to gain insight on cell metabolism and to provide an educational tool to communicate their findings through virtual reality experiences.

## 4.1. 3-D Network Visualization

Although most of the existing methods to draw metabolic pathways are in 2D space, there are a few 3D graph drawing algorithms, such as force-directed drawing and orthogonal drawing (Closson et al., 1999; Landgraf, 2001). However, these algorithms produce graphs, which are difficult to interpret in three dimensions due to edge-crossings and graph complexity. Figure 2 shows a three-dimensional force-directed layout called GEM (adapted from the tulip graph package at www.tulip.org).

## 4.2. User Interaction and Navigation

Three-dimensional graph layouts can be difficult to interpret. The biologist can choose any node as a focus node for a reaction of interest (ROI). An ROI is defined as all the reactions that the focus node participates in. The reactions are positioned evenly in the 3D region around the focus node. Figure 3 shows an example of a selected ROI.

Text-based information concerning the source, synonyms and other data is important in the study of metabolic pathways. The tablet PC provides a method to present a traditional 2-D desktop interface to the user. The tablet PC's GUI can provide an easily accessible method for the user to find and travel to a node of interest. In this GUI, the user can see tables of the nodes and edges present in the displayed graph. The user can interactively select nodes and edges and see the complete information on the node or edge on the GUI. The user can also select to have the nodes and/or edges colored to highlight them in the scene or can select

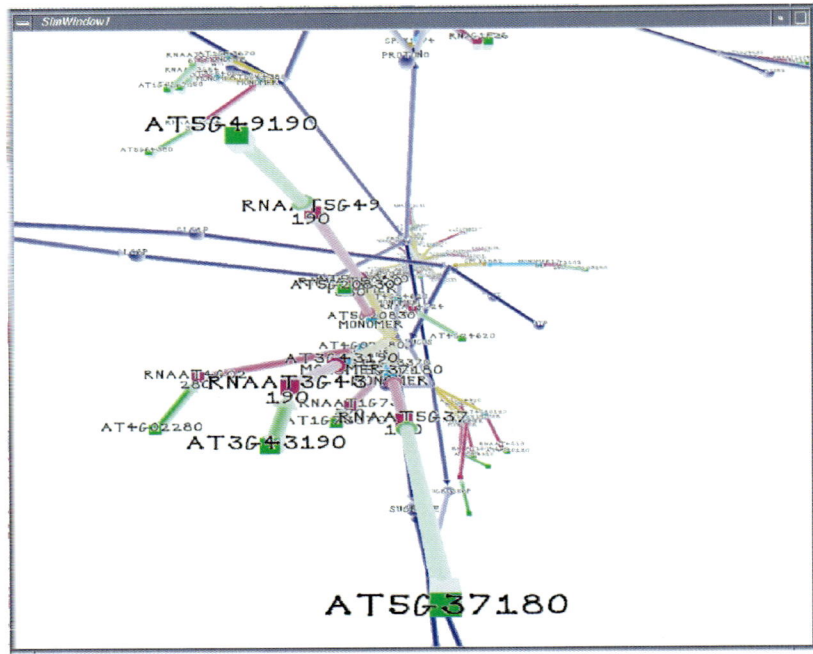

**FIGURE 2.** Partial graph of metabolic pathways in Arabidopsis. The graph layout uses a 3-D force-directed layout which puts highly connected nodes in the center of the graph.

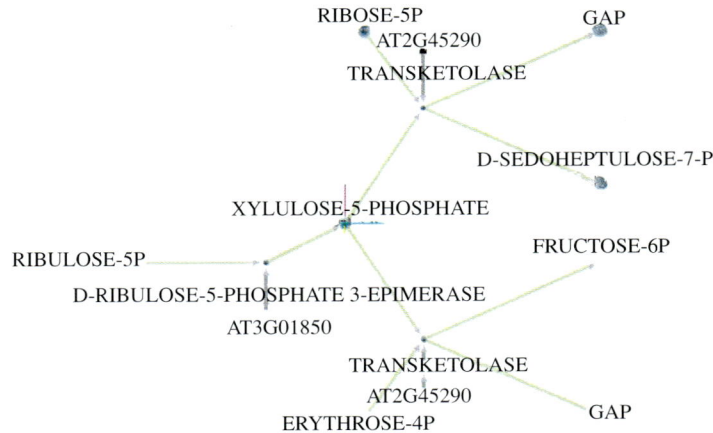

**FIGURE 3.** A reaction of interest selected from the Calvin Cycle. The focus node is xylulose-5-phosphate.

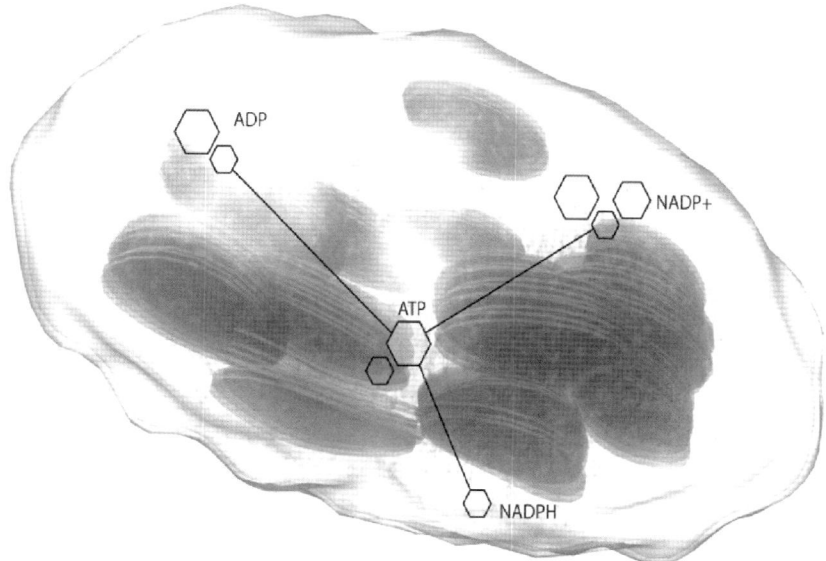

**FIGURE 4.** Three-dimensional model of the chloroplast with a network overlay showing one of the metabolic processes in that organelle.

a node or edge or have the VR environment moved to show the selected object. An example interaction is "pulling" a reaction-of-interest out from a larger graph. This tool uses Java for the GUI and a tool included in the VR software, Tweek (www.vrjuggler.org).

## 4.3. Virtual Cell Representation

As a teaching environment for high school and college students, we are developing a virtual cell model, and integrating this cell with the cellular metabolism and regulation models. The student will be able to visualize the cell from the outside, as well as cross-sections of the entire cell. As the student zooms inward, she/he enters the cell and the organelle systems within. From this organelle world, the student can track metabolic pathways, following anabolism and catabolism within the cell. This encompasses visualizing reactions within the organelle, and as a given metabolite leaves the cell, the student would virtually move from organelle to organelle. Figure 4 shows the virtual cell model with an overlay showing interactions.

## 5. SUMMARY

The FCModeler software is designed with a focus on understanding the complex molecular network in the model plant eukaryotic species, Arabidopsis.

FCModeler enables biologists to capture relationships at different levels of detail, to integrate gene expression data, and to model these relationships. Because of an absence of knowledge about many biological interactions, the software is designed to model at many levels of detail. The three-dimensional virtual environment offers exciting new opportunities for visualizing complex networks. The ability to link detailed physical models with representations of regulatory and metabolic flow will lead to new teaching methods in biology.

ACKNOWLEDGEMENTS. Funding for this project was provided by grants from the National Science Foundation in the Arabidopsis 2010 (DBI-0209809) and Information Technology Research (IBN-0219366) Programs. Seed funding was also provided by the Iowa State University Plant Sciences Institute and the Roy J. Carver Foundation. We thank Lucas Mueller and TAIR for helpful advice and for making the AraCyc pathways publicly available. Thank you to my Metabolic Networking project collaborators: Dr. Eve Wurtele, Dr. Dianne Cook, and Dr. Carolina Cruz-Neira. Thanks are also due to the dedicated group of students supporting the metabolic modeling efforts at Iowa State University: Pan Du, Yuting Yang, Joset Etzel, Jie Li, Adam Tonjack, Andres Reinot, and Kris Blom.

## 6. REFERENCES

Brown, M.P.S., Grundy, W.N., Lin, D., Cristianini, N., Sugnet, C.W., Furey, T.S., Ares, M., and Haussler, D., 2000, Knowledge-based analysis of microarray gene expression data by using support vector machines, *Proceedings National Academy of Science* **97**(1): 262–267.

Closson, M., Gartshore, S., Johansen, J., and Wismath, S.K., 1999, Fully Dynamic 3-Dimensional Orthogonal Graph Drawing, *7th International Symposium, GD'99, Stirín Castle, Czech Republic*, J. Kratochvíl. Berlin, Springer Verlag, **1731:** 49–58.

Cruz-Neira, C., Sandin, D.J., and DeFanti, T.A., 1993, *Surround-Screen Projections-Based Virtual Reality: The Design and Implementation of the CAVE*, ACM SIGGRAPH 93.

Dickerson, J.A., Cox, Z., Wurtele, E.S., and Fulmer, A.W., 2001, *Creating Metabolic and Regulatory Network Models using Fuzzy Cognitive Maps*, North American Fuzzy Information Processing Conference (NAFIPS), Vancouver, B.C.

Dickerson, J.A., Berleant, D., Cox, Z., Ashlock, D., Fulmer, A.W., and Wurtele, E.S., 2002, Creating and Modeling Metabolic and Regulatory Networks Using Text Mining and Fuzzy Expert Systems, *Computational Biology and Genome Informatics*, C.H. Wu, P. Wang and J.T.L. Wang. Hong Kong, World Scientific.

Dickerson, J.A., and Kosko, B., 1994, Virtual Worlds as Fuzzy Cognitive Maps, *Presence* **3**(2, Spring): 173–189.

Ding, J., Berleant, D., Nettleton, D., and Wurtele, E., 2002, *Mining MEDLINE: Abstracts, Sentences, or Phrases?*, Pacific Symposium on Biocomputing (PSB 2002), Kaua'i, Hawaii.

Dougherty, E.R., Barrera, J., Brun, M., Kim, S., Cesar, R.M., Chen, Y., Bittner, M., and Trent, J.M., 2002, Inference from clustering with application to gene-expression microarrays, *J Comput Biol* **9**: 105–126.

Fatland, B.F., Ke, J., Anderson, M., Mentzen, W., Cui, L.W., Allred, C., Johnston, J.L., Nikolau, B.J., and Wurtele, E.S., 2002, Molecular Characterization of a Novel Heteromeric ATP-Citrate Lyase that Generates Cytosolic Acetyl-CoA in Arabidopsis, *Plant Physiology* **130**: 740–756.

Hagiwara, M., 1992, *Extended Fuzzy Cognitive Maps*, 92 IEEE Int Conf Fuzzy Syst FUZZ-IEEE, San Diego, IEEE.

Kanehisa, M., and Goto, S., 2000, KEGG: Kyoto Encyclopedia of Genes and Genomes, *Nucleic Acids Research* **28**(1): 27–30.

Karp, P.D., Riley, M., Paley, S.M., and Pellegrini-Toole, A., 2002, The MetaCyc Database, *Nucl. Acids. Res.* **30**: 59–61.

Karp, P.D., Riley, M., Saier, M., Paulsen, I.T., Collado-Vides, J., Paley, S.M., Pellegrini-Toole, A., Bonavides, C., and Gama-Castro, S., 2002, The EcoCyc Database, *Nucl. Acids. Res.* **30**: 56–58.

Kosko, B., 1986, Fuzzy Cognitive Maps, *International Journal Man-Machine Studies* **24**: 65–75.

Landgraf, B., 2001, 3D Graph Drawing, *Drawing Graphs: Methods and Models*, D.W.M. Kaufmann. Berlin, Springer Verlag, **2025**: 172–192.

Lee, K., Kim, S., and Sakawa, M., 1996, On-line fault diagnosis by using fuzzy cognitive maps, *IEICE Transactions on Fundamentals of Electronics, Communications and Computer Sciences* **E79-A,**(6): 921–922.

Oliver, D.J., Nikolau, B.J., and Wurtele, E.S., 2002, Functional Genomics: High throughput mRNA, protein, and metabolite analyses, *Metabolic Engineering* **In Press**.

Overbeek, R., Larsen, N., Pusch, G.D., D'Souza, M., Jr, E.S., Kyrpides, N., Fonstein, M., Maltsev, N., and Selkov, E., 2000, WIT: integrated system for high-throughput genome sequence analysis and metabolic reconstruction, *Nucl. Acids. Res.* **28**: 123–125.

Yao, T., 2002, Bioinformatics for the genomic sciences and towards systems biology. Japanese activities in the post-genome era, *Prog Biophys Mol Biol* **80**: 23–42.

*Chapter 7*

# Old Methods for New Ideas: Genetic Dissection of the Determinants of Gene Expression Levels

Kyunga Kim, Marilyn A.L. West, Richard W. Michelmore, Dina A. St. Clair, and R.W. Doerge

## 1. INTRODUCTION

There is increasing interest in understanding the molecular basis of complex traits. Initially, the genetic dissection of quantitative traits involved measurements of gross phenotypes. Subsequently, specific physiological and developmental components of individual traits have been dissected. Most recently, the underlying mechanisms of inheritance have been studied through various approaches that are supported by modern technological and methodological advances, namely quantitative trait locus/loci (QTL) analysis (Mackay, 2001; Mauricio, 2001; Doerge, 2002) and mutant analysis (Rossant and Spence, 1998; Hughes et al., 2000) in genetics; genome sequencing (Jang et al., 1999; The Arabidopsis Genome Initiative, 2000; Mouse Genome Sequencing Consortium, 2002) and gene expression analysis (Duggan et al., 1999; Lipshutz et al., 1999) in genomics; and protein structure analysis (Service, 1999) and protein assay (Kodadek, 2001; MacBeath, 2002)

**Kyunga Kim**   Department of Statistics, Purdue University, West Lafayette, Indiana 47907; **Marilyn A.L. West, Richard W. Michelmore, and Dina A. St. Clair**   Department of Plant Sciences, University of California Davis, Davis California, 95616;   **R.W. Doerge**   Department of Statistics and Department of Agronomy, Purdue University, West Lafayette, Indiana 47907

*Genome Exploitation: Data Mining the Genome*, edited by J. Perry Gustafson, Randy Shoemaker, and John W. Snape.
Springer Science + Business Media, New York, 2005.

in proteomics. Since each technology and approach focuses on specific pieces of the larger, poorly understood systems biology, the challenge is to integrate these different types of information to elucidate the genetic architecture of complex traits. In particular, the regulation of complex traits remains poorly understood, and there are still large gaps in our understanding of regulatory networks.

Statistically, QTL analysis has offered many interesting theoretical challenges and complex models that have resulted in useful software. The conclusions of QTL analysis often point to large regions of the genome, typically containing many genes, being associated with a measured quantitative (phenotypic) trait of interest. These QTL are largely regions of unknown function that often disappear in the next experiment or environment. If QTL are localized, and a small number of candidate genes established, it requires large populations of recombinants and extensive replicated experimentation. The focus now is to move beyond the association of molecular markers with quantitative phenotypes to understand the regulation of gene expression and its consequences on the variation of quantitative traits. To achieve this goal, more powerful statistical methods are needed to reveal the genes controlling the expression of complex traits. Proper experimental design and the application of appropriate statistical methodologies to gene expression levels will provide insights into regulatory networks controlling transcript levels and ultimately the regulation of complex trait phenotypes.

## 2. GENETICAL GENOMICS

In recent years, there has been growing interest in uniting genetic and genomic approaches to enable more comprehensive dissections of complex traits and their genetic architecture. Jansen and Nap (2001) termed this synthesis 'genetical genomics' and Doerge (2002) outlined the use of QTL methodology to analyze gene expression data from microarray experiments. Unraveling the mechanisms of phenotypic control and the determinants of variation found in gene expression are now the main foci of many investigations. Several groups have attempted to integrate quantitative genetic analysis and gene expression analysis. Jin et al. (2001) used a mixed model ANOVA approach to demonstrate significant genotypic factors in *D. melanogaster* affecting variation in gene expression levels, along with other factors such as sex and age. Wayne and McIntyre (2002) complemented a QTL fine mapping study of *D. melanogaster* with a follow-up microarray experiment to investigate candidate genes that were likely involved in controlling ovariole number. While their approach does not reveal causal relationships, it is a creative avenue to identify candidate genes associated with complex traits. Brem et al. (2002) applied genetic mapping to microarray data for the detection of the genomic regions in yeast affecting gene expression levels. These investigations employed simple statistical tools (e.g., the Wilcoxon-Mann-Whitney test, $X^2$ test, ANOVA) for single marker analyses based on very small populations (6-40 individuals), and thus are limited in their statistical power to determine the relationship between genotypic and phenotypic data.

Recently, Schadt et al. (2003) used QTL interval mapping (Lander and Botstein, 1989) on gene expression data from human, maize and mouse experiments. In one case, they analyzed 111 individuals from an $F_2$ mouse population derived from two standard inbred strains. Oligonucleotide microarrays were employed to evaluate the expression of 23,574 genes in both the parental lines and the 111 $F_2$ individuals; 7,861 differentially expressed genes were identified within each of the parental lines or in at least a tenth of the 111 $F_2$ individuals. Treating the expression levels of the 7,861 genes as quantitative traits, interval mapping using MAPMAKER/QTL (Lincoln et al., 1992) was employed to identify genomic regions (expression-QTL, or e-QTL) associated with gene expression variation.

The use of QTL mapping methods on microarray gene expression data outlined by Doerge (2002) and now referred to as e-QTL (expression quantitative trait locus) mapping (Schadt et al., 2003) aims to identify the determinants of polymorphisms in gene expression levels (expression level polymorphisms or ELPs) through genetic analysis. These existing approaches have the potential to provide insights into regulatory networks controlling complex phenotypes, but may be limited in their statistical power. Since no one has yet investigated the statistical power of existing QTL methodologies within the context of e-QTL mapping, there may be several statistical and technological issues that need to be adequately addressed.

We are therefore investigating whether the application of an existing QTL methodology known as multiple trait QTL mapping is adequate for e-QTL mapping, or whether new statistical methodologies need to be developed. Multiple-trait QTL analysis was developed by Jiang and Zeng (1995) to provide statistical tests for gene-by-environment interactions, as well as genomic regions associated with the multiple quantitative traits. It may be possible to exploit the methods provided by Jiang and Zeng within the framework of gene expression data for the purpose of locating e-QTL that are putative determinants of expression level polymorphisms. In our analysis we treat gene expression levels under different experimental conditions as phenotypes representing multiple traits in the same manner that Jiang and Zeng (1995) treated phenotype measurements under different environmental conditions as multiple traits.

## 3. METHODS

The application of appropriate statistical methodology to gene expression data from a segregating population allows putative determinants of ELPs to be mapped as e-QTLs to specific genomic regions by using a partitioning (decomposition) of the sources of variation associated with measured transcript levels (e.g., genetic and non-genetic factors, and their interactions). Differences in gene expression (ELPs) may be determined by cis factors associated with the gene exhibiting the ELP, or by trans factors encoded elsewhere in the genome (regulatory loci). We define a structural locus as the genomic region that contains a gene whose transcript is measured under the different experimental conditions and a regulatory locus as an independent gene that controls the expression of the structural gene. Both structural and regulatory loci can be determinants of ELPs since both contribute to gene expression levels. Therefore, both structural and regulatory loci can be identified as e-QTLs. Analysis of the genomic locations of e-QTLs relative to the genes exhibiting ELPs allows the distinction of regulatory from structural loci, thus providing insights into regulatory networks underlying gene expression and ultimately the genetic basis of complex traits encoded by these genes.

### 3.1.  Dye Swap Experimental Design

Genes are represented as probes (cDNAs or oligonucleotides) on a microarray by arraying them on a solid substrate. The amount of sample hybridized to each spot or gene feature on the array is quantified reflecting the amount of transcript for each gene (Schena et al., 1996). Various statistical methods (Eisen et al., 1998; Holter et al., 2000; Kerr et al., 2000; Kerr and Churchill, 2001a,b,c; Newton et al., 2001) have been employed to identify statistically significant changes in gene expression. A variety of statistical issues associated with microarray analysis have been raised that have not been fully resolved and are the continuing focus of investigations to assess differential gene expression. Regardless of the limitations of current methodologies, the end results of microarray analyses do not delineate causal relationships, but only indicate gene expression changes in response to some stimulus or change in experimental conditions.

We have employed the dye swap experimental design that has been extensively used in microarray experiments and the data it produces to locate determinants of variation in gene expression on a genetic map. Gene expression is typically measured in response to a treatment condition and compared to a control condition, or to another treatment condition. These comparisons are often made between pairs of conditions, and in a dye swap design are based upon a two dye (Cy3:green and Cy5:red) labeling system. mRNA taken from samples of tissue under different experimental conditions are hybridized to a microarray to identify gene expression differences. For each gene, the amount of hybridization is measured by examining the fluorescence frequencies of the two dyes. Because significant dye-specific labeling artifacts have been observed, a second hybridization is often performed in which the biological samples are the same, but the labeling with the dyes are reversed (i.e., "swapped"). Such a dye swap allows the researcher to account for the bias due to dye labeling either by incorporating the correction into a pre-processing normalization, or by incorporating the dye effect as a source of additional variation in the statistical analysis.

### 3.2  Modelling Determinants of Expression Level Polymorphisms

The genetic architecture of the regulatory networks controlling ELPs is potentially very complex. Three simple models that are based on a single gene with one or two regulatory loci are shown in Figure 1. The simplest scenario that is used later as the basis of our investigation (Model 1; Figure 1) contains only a single structural gene whose variation is determined by a single regulatory gene. In the

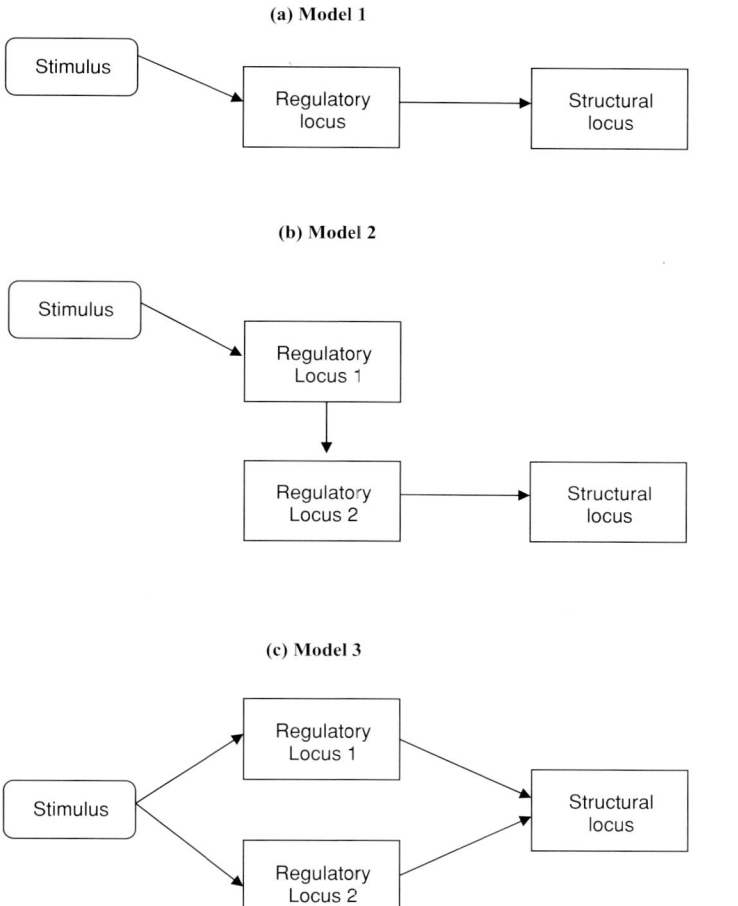

**FIGURE 1.** Three simple models illustrating a single structural gene with one or two additional determinants of expression level polymorphisms.

proposed models (Figure 1), the regulatory genes respond to a stimulus and their products interact with the structural gene to promote or repress its expression. Assuming there is some allelic variation in a segregating population or among genetically distinct individuals contributing to polymorphic gene expression, one can partition (decompose) the observed variation in gene expression into several components: genetic sources such as sequence polymorphism (allelic differences) at both structural and regulatory loci; non-genetic sources such as treatment effects (the nature of the stimulus); interactions between genetic and non-genetic components (genotype and the stimulus); and also systematic or technological components of the experiment itself, such as dye and array effects.

We have used a well-established statistical method, linear models (Searle, 1971), to partition the sources of variation contributing to differences in gene expression. Linear models have been applied extensively in both QTL mapping (Zeng, 1993; Jiang and Zeng, 1995) and microarray analysis (Kerr et al., 2000; Kerr and Churchill, 2001a). Our investigation here is focused on a simple linear genetic model that includes a single structural gene and a single regulatory locus (Model 1; Figure 1), but it is possible to extend this model to more complex models (e.g., Models 2 and 3, Figure 1). For an additive model (i.e., no epistasis), we employ a segregating recombinant inbred line (RIL) population (see section 3.3), and for every individual RIL let $i$ denote the genotype of the regulatory locus, j denote the genotype of the structural locus, and let $y_{ijklmr}$ denote the gene expression measurement of the structural gene as measured by spotted microarray technology. The measurement of gene expression is in either the original, or log scale, and obtained under treatment $k$ using dye $l$ on array $m$ in replication $r$:

$$y_{ijklmr} = \mu + \alpha_i + \beta_j + \tau_k + (\alpha\tau)_{ik} + \delta_l + A_m + \varepsilon_{ijklmr} \tag{1}$$

where $i = 1, 2, j = 1, 2, k = 1, 2, l = 1, 2, m = 1, 2$, and $r = 1, \ldots, R$. The terms $\alpha$, $\beta$, $\tau$ and $\delta$ are the additive effects of the regulatory and structural loci, treatment effect, and dye effect, respectively. The interaction between the regulatory loci and the treatment is denoted $\alpha\tau$, while $A$, the array effect, is assumed to be distributed as a random normal with mean 0 and variance $\sigma_A^2$. The measurement error $\varepsilon_{ijklmr}$ is distributed as a random normal with mean 0 and variance $\sigma_\varepsilon^2$. The array effect and measurement error are assumed to be independent. It should be noted that in this simple model we did not consider statistical interactions between the treatment and the structural locus.

### 3.3. Experimental Design for Expression Level Polymorphism Analysis

Variation in gene expression between genetically distinct individuals, i.e. biological variation, is not usually assessed in microarray experiments. Instead, the genotypes are assumed to be uniform (Black and Doerge, 2002; Churchill, 2002), and differential expression is tested. In fact, variation in gene expression between individuals of a segregating population can be assessed and several studies have already analyzed genetic/allelic variation in gene expression (Cowles et al., 2002; Oleksiak et al., 2002; Yan et al., 2002a,b). Several types of segregating populations have been proposed for mapping e-QTL; these include recombination congenic strains (RCSs), recombination inbred lines (RILs), and chromosome substitution strains (CSSs) (Jansen, 2003). These types of populations provide homogeneous families that allow the biological replication that is essential for distinguishing genetic from non-genetic sources of variation. For example, individuals within a homozygous RIL are genetically identical, thus phenotypic variation within a RIL should be due to non-genetic sources, whereas variation among RILs is due to both genetic and non-genetic sources.

For many species, RILs are the easiest of the previously mentioned types of population to generate. RILs derived from divergent inbred parents would be

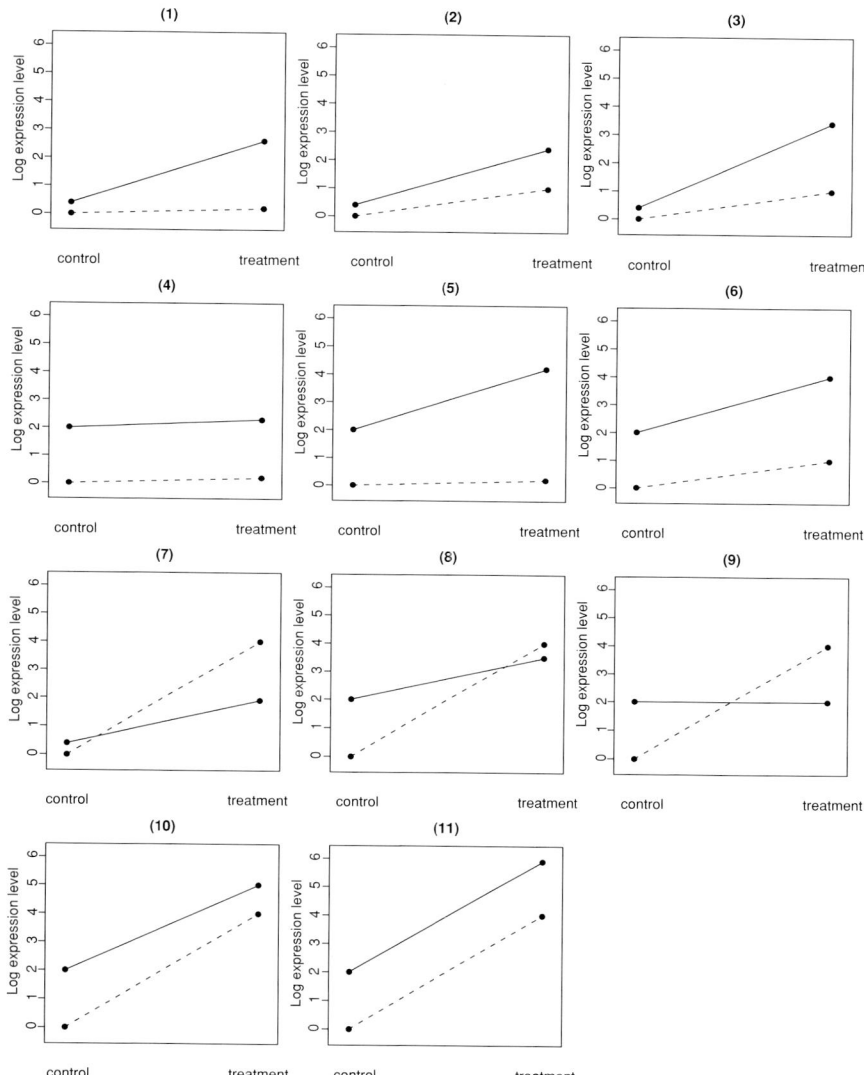

**FIGURE 2.** Examples of gene expression profiles with potential expression level polymorphisms. The dashed lines indicate one parental line (reference line); solid lines indicate the other parental line. The vertical axis denotes (relative) expression levels where the expression level for the reference line with no treatment (control) was adjusted to 0. The cutoff value for potential expression level polymorphisms was assumed to be adjusted to 2-fold change.

expected to segregate for ELPs for a subset of genes controlling a complex trait. Examples of potential types of ELPs are illustrated in Figure 2. A preliminary genome-wide expression analysis can be conducted on potential inbred parent lines to identify lines that exhibit maximum levels of ELPs under the experimental conditions being investigated (Figure 2). The lines showing the largest number of potential ELPs can be used as parents to derive a segregating RIL population. The expectation is that some, but not necessarily all, of the potential ELPs will supply information that will allow e-QTL mapping.

Each individual in the segregating (RIL) population is genotyped with DNA markers. The RIL population is then subjected to replicated experiments under specific experimental conditions and treatments, and the expression profile of each RIL is determined using microarrays. In order to properly assess the variation in gene expression, ELP mapping requires multiple microarrays for every RIL in the segregating population. At least two biological replicates per RIL for every treatment in the experiment are required.

## 3.4. Implementation of ELP Mapping Using Existing Multiple Trait Mapping Methodology

Because the expression of each gene is measured for each individual over different treatment conditions, gene expression measurements can be viewed as multiple evaluations of a trait in different environments. Multiple trait analysis (Jiang and Zeng, 1995) is an extension of single trait QTL analysis to multiple trait QTL analysis using composite interval mapping (CIM) (Zeng, 1993, 1994). Within the context of traditional QTL mapping each individual's phenotype is denoted by the linear model, $w_p$, where the subscript $p$ distinguishes among the multiple traits,

$$w_p = b_{0p} + b_p^* x^* + \sum_{l=1}^{t} b_{lp} x_l + e_p \quad (p = 1, \ldots, P), \qquad (2)$$

and where $b_{0p}$ is the overall mean for trait $p$; $b_p^*$ is the additive effect of the putative QTL on trait $p$; $x^*$ is the number of alleles at the putative QTL; $b_{lp}$ is the partial regression coefficient of $w_p$ on $x_l$; $x_l$ is the allele of marker $l$ (among $t$ markers used for controlling residual genetic variation) on the individual; and $e_p$ is the residual effect on trait $l$ for the individual. The residuals are correlated among the $P$ traits within individuals while they are independent among genetically distinct individuals. Since multiple trait CIM accounts for the correlation structure among the traits, it provides estimates of location for the genomic regions or QTL that are associated with the multiple traits. Multiple trait CIM accommodates our proposed setting and linear additive genetic model (1) for a simple biological model (Model 1; Figure 1), including statistical interactions between the treatment (stimulus) and the *regulatory* loci. Jiang and Zeng (1995) suggest that the multiple trait CIM approach can be used to test the gene-by-environmental interactions when the same trait is assessed in different environmental conditions. Based upon this, we tested the application of multiple trait analysis for ELP experiments based on a dye swap microarray experiment for expression of a structural gene measured under two treatments.

This experimental design provides four "traits" that can be described using equation (1), and can be extended to include more complicated models. The four traits are the control and treatment measurements of a single gene's transcript under a two dye swap design, and are represented as follows:

$$w_1 = \frac{1}{R} \sum_{r=1}^{R} y_{ij111r} = \mu + \alpha_i + \beta_j + \tau_1 + (\alpha\tau)_{i1} + \delta_1 + A_1 + \frac{1}{R} \sum_{r=1}^{R} \varepsilon_{ij111r}$$

$$w_2 = \frac{1}{R} \sum_{r=1}^{R} y_{ij221r} = \mu + \alpha_i + \beta_j + \tau_2 + (\alpha\tau)_{i2} + \delta_2 + A_1 + \frac{1}{R} \sum_{r=1}^{R} \varepsilon_{ij221r}$$

$$w_3 = \frac{1}{R} \sum_{r=1}^{R} y_{ij122r} = \mu + \alpha_i + \beta_j + \tau_1 + (\alpha\tau)_{i1} + \delta_2 + A_2 + \frac{1}{R} \sum_{r=1}^{R} \varepsilon_{ij122r}$$

$$w_4 = \frac{1}{R} \sum_{r=1}^{R} y_{ij212r} = \mu + \alpha_i + \beta_j + \tau_2 + (\alpha\tau)_{i2} + \delta_1 + A_2 + \frac{1}{R} \sum_{r=1}^{R} \varepsilon_{ij212r}$$

$$(3)$$

## 4. SIMULATION STUDY

For many complex statistical methodologies, including QTL analysis, it is difficult to analytically assess performance (i.e., how well the methods accomplish what they are created to do). Simulation studies provide a way to evaluate the performance of such methodologies before actual biological experiments are conducted (Broman and Speed, 2002). Based on the simplest biological model (Model 1; Figure 1) data were simulated and used to investigate whether existing multiple trait QTL mapping methodologies can be used to identify determinants of expression level polymorphisms. We employed the *JZmapqtl* procedure (Basten et al., 1994) in QTL-Cartographer (Basten et al., 2002), that was designed for multiple trait composite interval mapping of traditional QTLs, to identify chromosomal regions (e-QTL) that affect, control, and/or determine the expression of structural genes (i.e., ELPs).

### 4.1. Simulation of RIL Genotypes

We simulated RIL populations consisting of 100, 200, 300, 400, 500, 700, and 1000 progeny for the purpose of assessing the sample size required for e-QTL detection, given the remainder of the parameters that characterize the genome. Using the Arabidopsis genome as our model, we considered five chromosomes with lengths of 135, 100, 100, 125, and 140 cM, and 120 markers were equally distributed at 5 cM intervals over the genome. Two unlinked e-QTL were simulated with equal additive effects of either 0.10 or 0.50. The regulatory locus was placed 98 cM from the top end of chromosome 1 and the structural locus was located 27 cM from the top end of chromosome 2. The remaining chromosomes did not contain any e-QTL and epistasis was not considered. Each structural and regulatory locus, as well as each marker, was assumed to be biallelic with equal allele frequencies of 0.50. Genotypes of each marker linked to the e-QTLs and all other markers on the five chromosomes were randomly simulated and assigned according to their allele frequencies. We employed Kosambi's map function assuming recombination with moderate interference. The genotype of each linked marker on chromosomes 1 and 2 was assigned based on the recombination fraction and genotype of the corresponding e-QTL.

## 4.2. Simulation of Gene Expression Simulation

For each individual RIL in the population, the gene expression measurements were simulated based on the proposed mixed linear genetic model

$$y_{ijklm} = \mu + \alpha_i + \beta_j + \tau_k + (\alpha\tau)_{ik} + \delta_l + A_m + \varepsilon_{ijklm} \tag{4}$$

where $i = 1, 2, j = 1, 2, k = 1, 2, l = 1, 2$, and $m = 1, 2$; $y_{ijklm}$ is the gene expression level in original or log scale assessed under treatment $k$ with dye $l$ on array $m$; $\alpha_i, \beta_j$ are the additive effects of the *regulatory* and the *structural* loci; $i$ and $j$ correspond to the simulated genotypes; $\tau$ and $\delta$ are the effects of the treatment, and dye, respectively; $A$ is a random normal array effect with mean 0 and variance $\sigma_A^2 = 1$; and $\varepsilon_{ijklm}$ is the measurement error distributed as a random normal with mean 0 and variance $\sigma_\varepsilon^2 = 1$. The array effect and measurement error are assumed to be independent, and a range of parameter values (Table 1) for various examples of potential ELPs were considered (corresponding to Figure 2) in our study.

### Table 1
**Parameter Configurations (1 Through 11 Refer to Figure 2) and Heritabilities of Four Genes/Traits. The *Regulatory* e-QTL was Simulated 98cM from the Top End of Chromosome 1; the *Structural* Locus was Simulated 27cM from the Top End of Chromosome 2. The Dye Effect was Simulated with a Value of 0.20; and the Variance of Both the Array Effect and the Measurement Error was 1. The Values of the Interaction were Set to Satisfy: $(\alpha\tau)_{11} + (\alpha\tau)_{12} + (\alpha\tau)_{21} + (\alpha\tau)_{22} = 0$ and $(\alpha\tau)_{11} = (\alpha\tau)_{12} = (\alpha\tau)_{21}$.**

| Config. | Additive Effect | | Trmt. Effect | Interact. | Heritability | |
|---|---|---|---|---|---|---|
| | Reg. | Struc. | | | Trait 1,3 | Trait 2,4 |
| (1) | 0.1 | 0.1 | 0.1 | 1.500 | 0.01 | 0.38 |
| (2) | 0.1 | 0.1 | 0.5 | 0.750 | 0.01 | 0.16 |
| (3) | 0.1 | 0.1 | 0.5 | 1.500 | 0.01 | 0.38 |
| (4) | 0.5 | 0.5 | 0.1 | 0.075 | 0.20 | 0.22 |
| (5) | 0.5 | 0.5 | 0.1 | 1.500 | 0.20 | 0.56 |
| (6) | 0.5 | 0.5 | 0.5 | 0.750 | 0.20 | 0.38 |
| (7) | 0.1 | 0.1 | 2.0 | −1.875 | 0.01 | 0.40 |
| (8) | 0.5 | 0.5 | 2.0 | −1.875 | 0.20 | 0.29 |
| (9) | 0.5 | 0.5 | 2.0 | −3.000 | 0.20 | 0.56 |
| (10) | 0.5 | 0.5 | 2.0 | −0.750 | 0.20 | 0.11 |
| (11) | 0.5 | 0.5 | 2.0 | −0.075 | 0.20 | 0.18 |

## 4.3. Multiple Gene/trait JZmapqtl Mapping

*JZmaptqtl* analysis was employed to analyze the simulated gene expression data from the RILs, each with a *regulatory* genotype $i$ and a *structural* genotype $j$ (Model 1; Figure 1). Four genes/traits were considered in a multiple trait setting for all RILs in the population. The four states of a gene's expression were the result of two treatment conditions ($\tau$), and each of these samples were labelled with each of the two dyes ($\delta$). Specifically, four gene expression measurements can be detailed as follows:

$$w_1 = y_{ij111} = \mu + \alpha_i + \beta_j + \tau_1 + (\alpha\tau)_{i1} + \delta_1 + A_1 + \varepsilon_{ij111}$$
$$w_2 = y_{ij221} = \mu + \alpha_i + \beta_j + \tau_2 + (\alpha\tau)_{i2} + \delta_2 + A_1 + \varepsilon_{ij221}$$
$$w_3 = y_{ij122} = \mu + \alpha_i + \beta_j + \tau_1 + (\alpha\tau)_{i1} + \delta_2 + A_2 + \varepsilon_{ij122}$$
$$w_4 = y_{ij212} = \mu + \alpha_i + \beta_j + \tau_2 + (\alpha\tau)_{i2} + \delta_1 + A_2 + \varepsilon_{ij212} \qquad (5)$$

Heritabilities of the four traits were calculated under the various parameter configurations (Table 1). The simulated data were then analyzed by *JZmapqtl* procedure in QTL-Cartographer version 1.61v (Basten et al., 2002). Model 6 was employed with a walking speed of 1 cM and a window size of 10 cM. Among the results of *JZmapqtl* is the assessment of the joint likelihood that there is no joint additive QTL effect on the four traits/genes, and that there is no QTL by environment interaction. This first hypothesis is the equivalent of testing for significant e-QTL (both regulatory and structural), while the second hypothesis tests whether there is a significant interaction between the e-QTL and the treatment. In the latter situation, variation in the expression of the structural gene may not be statistically significant, but its expression may still be controlled by the regulatory gene.

Separate empirical thresholds based on 1000 permutations (Churchill and Doerge, 1994) were estimated independent of QTL-Cartographer for both tests, namely detecting significant e-QTL and significant interaction with between the treatment and determinants of ELPs. When estimating permutation thresholds for multiple genes/traits, it is essential that the randomizations maintain the correlation structure between genes/traits within each individual.

Once significant e-QTLs were determined for each gene via *JZmapqtl*, the resulting e-QTLs were compared to the actual genetic map. Because the biological model that was used for the simulations (only Model 1; Figure 1 reported here) was simple, *JZmapqtl* typically produced the same number of e-QTL as were delineated in the simulation model. As the simulation model or true biological system becomes more complex, the false positive rate for e-QTLs detected by *JZmapQTL* is likely to increase.

The effective number of individuals required for a segregating population of RILs was studied across 200 (repeated) data sets, which were repeatedly simulated under each parameter configuration and with each population size. The statistical power to detect and locate determinants of ELPs was calculated based on the proportion of multiple gene/trait analyses that significantly identified and located e-QTL within a 1 cM neighborhood of their true locations. Furthermore,

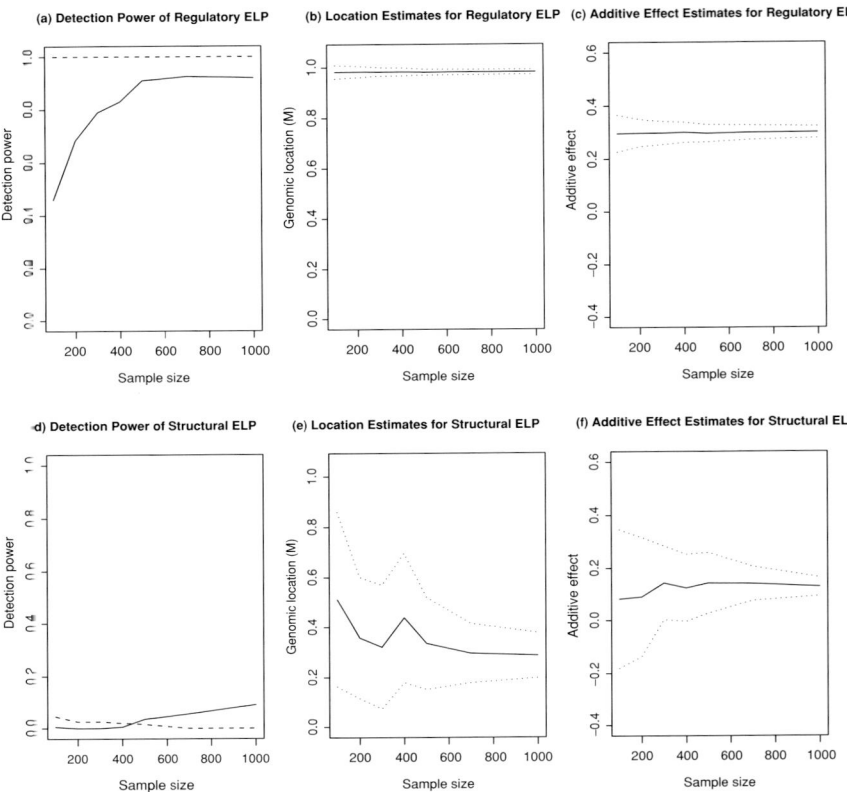

**FIGURE 3.** Statistical power and estimates with respect to sample sizes 100, 200, 300, 400, 500, 700 and 1000 based on 200 simulation runs for configuration 1. The *regulatory* locus was simulated 98cM from the top end of chromosome 1 having additive effect of 0.10 and treatment interaction; the *structural* locus was simulated 27cM from the top end of chromosome 2 with additive effect of 0.10 and no treatment interaction. For (a), (d) the solid lines represent detection power for e-QTL; dashed lines represent the detection power for interaction between e-QTL and treatment. For (b), (c), (e), (f) the solid lines represent the sample mean of the estimates of location and additive effect; dotted lines represent the respective 1-standard error limits.

the statistical power to detect the interaction between the e-QTL and the treatment was estimated by examining how many times the interaction was significantly detected out of 200 simulation repetitions. The sample means and standard deviations over all 200 runs for the estimates of e-QTL positions and additive effects were computed for further exploration of the properties of multiple trait CIM when applied to gene expression data measured over a population of individuals. The use of 200 runs for each simulation setting was determined by evaluating the results of simulations conducted under 100, 200, 300, and 500 runs. The effective gain in information between 200 runs and 500 runs was minimal, while the effective savings of computational resources was significant.

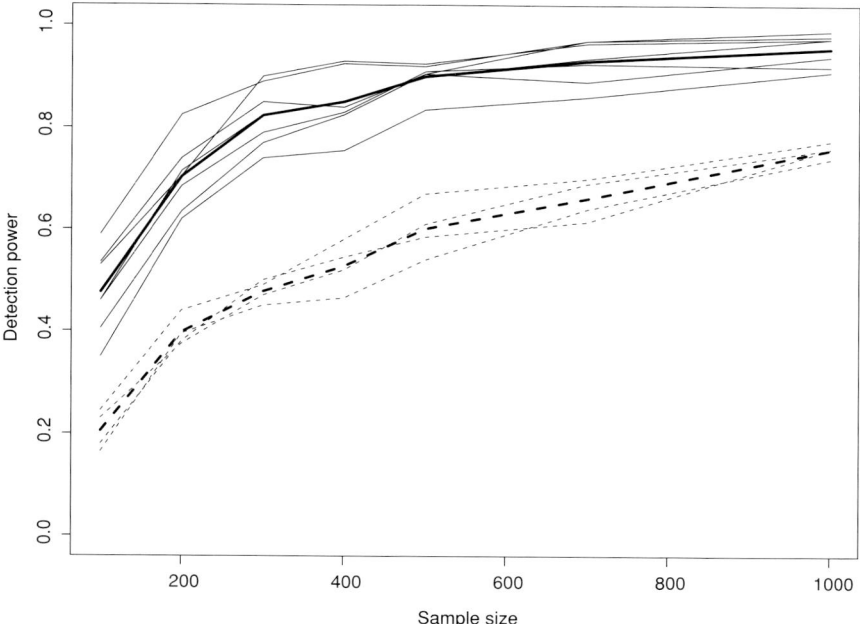

**FIGURE 4.** Detection power for location of regulatory locus with respect to sample sizes 100, 200, 300, 400, 500, 700, and 1000. The lines represent detection power for the *regulatory* locus under each configuration. Solid lines denote relatively high power; dashed lines denote low power. Bold lines (solid and dashed) represent the average power over the corresponding high and low power configurations.

## 4.4. Simulation Results

We present simulation results (Figures 3–5) for Model 1 (Figure 1) in our investigation of *JZmaptqtl* to detect and identify e-QTL. The estimated power for identifying the *regulatory* locus was higher than the power to detect the *structural* locus, although their additive effects are same (0.10). This is most likely due to the interaction between the regulatory gene and the treatment that was incorporated into the linear additive model and the lack of inclusion of a similar interaction term between the structural gene and the treatment. The standard deviation of the estimates were much smaller for the *regulatory* locus than for *structural* locus, and the sample bias of the additive effect is relatively large (0.20) for the *regulatory* locus, relative to the *structural* locus (0.02). Both the parameter estimates and the statistical power of the remaining parameter configurations (Table 1) supported the conclusion that the precision of the estimates for the regulatory locus is better than that for the structural locus (Table 2) under the conditions of our model. However, because there was no interaction term between the treatment and the structural locus, it is unclear whether this conclusion can be generalized to more complex situations.

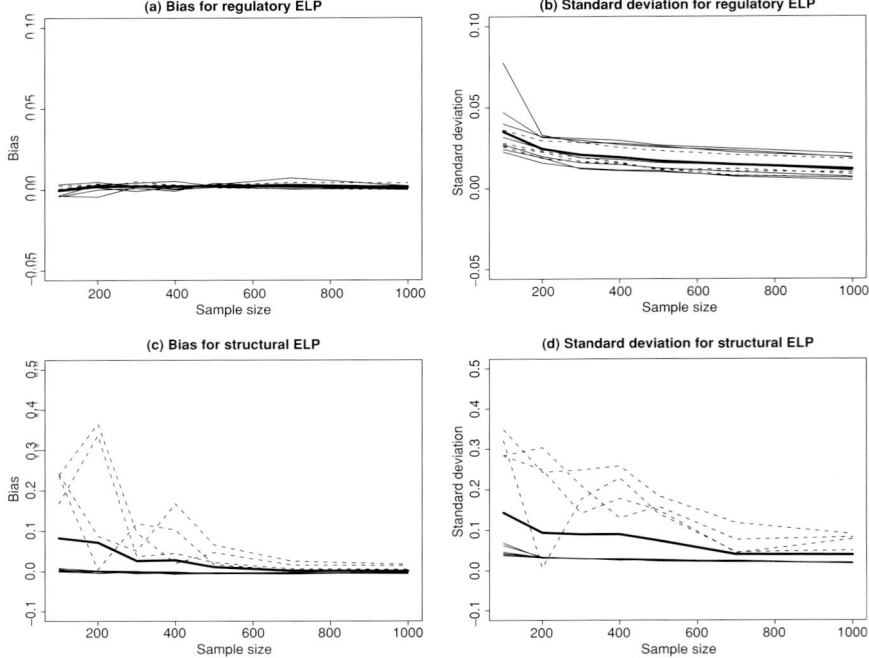

**FIGURE 5.** Bias and standard deviation of location estimates with respect to sample size 100, 200, 300, 400, 500, 700, and 1000. The narrow lines denote each configuration; bold lines denote the average values over all configurations. Solid lines denote configurations with additive effect = 0.50; dashed lines are for those with additive effect = 0.10.

**Table 2**

**Sample Means Corresponding to Estimates (Standard Error in Parentheses) over 200 Simulation Runs. Power is the Estimated Detection Power for e-QTL and Statistical Interaction (in Parenthesis), and the Additive Effects are Denoted as Add. The Configurations (C) 1–11 Correspond to Table 1.**

| (C) | Regulatory | | | Structural | | |
|---|---|---|---|---|---|---|
| | Loc. (cM) | Add. | Power | Loc. (cM) | Add. | Power |
| (1) | 98.23 (0.99) | .30 (.022) | .92 (1) | 28.81 (8.98) | .13 (.035) | .09 (0) |
| (2) | 98.41 (1.80) | .28 (.023) | .78 (1) | 28.49 (8.24) | .13 (.038) | .11 (.005) |
| (3) | 98.17 (0.87) | .30 (.024) | .94 (1) | 26.97 (5.00) | .13 (.023) | .09 (.005) |
| (4) | 98.19 (1.92) | .52 (.028) | .76 (.04) | 26.76 (1.78) | .50 (.027) | .78 (0) |
| (5) | 98.06 (0.65) | .50 (.029) | .98 (1) | 26.43 (1.94) | .50 (.027) | .73 (.005) |
| (6) | 98.24 (1.09) | .60 (.030) | .91 (1) | 26.63 (1.75) | .50 (.028) | .80 (.005) |
| (7) | 98.00 (0.65) | −.21 (.020) | .98 (1) | 27.56 (7.76) | .13 (.048) | .12 (.005) |
| (8) | 98.07 (0.71) | −.05 (.019) | .98 (1) | 26.84 (1.97) | .50 (.028) | .78 (0) |
| (9) | 98.08 (0.50) | −.10 (.015) | .99 (1) | 26.55 (1.87) | .50 (.026) | .76 (.005) |
| (10) | 98.23 (1.89) | .20 (.027) | .76 (1) | 26.64 (2.03) | .51 (.026) | .72 (0) |
| (11) | 97.99 (2.13) | .47 (.028) | .74 (.02) | 26.82 (1.88) | .50 (.030) | .75 (.005) |

The estimates of location and additive effects seem relatively unbiased except for additive effects of the *regulatory* locus (Figure 3). As noted previously, this bias may have been caused by specific effects in statistical model, such as the treatment and the treatment by regulatory locus interaction. Despite the biased estimation of the additive effect, e-QTL mapping via *JZmapQTL* detected *regulatory* locus fairly well with good identification of the treatment interaction; however it showed poor power to detect the *structural* locus, especially those with small effects. The relative gain that is illustrated for regulatory locus location and effect over the structural loci is most likely the result of the biological model that was used as the basis for the simulation study. This biological model (Model 1), while simple (1 structural locus, 1 regulatory locus), reflected a statistical interaction between the treatment and the regulatory locus that the structural locus did not undergo. The interaction between the regulatory locus and a treatment reflects that a determinant of an ELP can be distinct from the transcript that is being regulated and may not encode the mRNA that is changing.

Because a primary goal of ELP analysis is to identify regulatory loci at their correct locations, we focused on the statistical power to locate regulatory e-QTL under various configurations (Model 1; Table 1; Figure 2). The results of this investigation divided the parameter configurations into two distinct groups based on their statistical power. The high-power group includes configurations (1), (3), (5), (6), (7), (8), and (9); the low-power group includes configurations (2), (4), (10), and (11) (Figure 4; configurations refer to gene expression profiles in Figure 2). Specifics of the low-power group are that the interaction is small and trait heritabilities of all genes/traits are less than 0.25. Interestingly, since the *JZmapqtl* analysis, as applied to gene expression data, accounts for the statistical interaction, it appears to some degree that it boosts the power to locate e-QTL that interact with the treatment; however, the gain in power becomes less after a sample size of 500 is achieved. By examining the increasing pattern of the power, we found that the continued gain in power becomes less after the sample size of 500 is achieved. For a sample size of 500, the average power to locate e-QTL is 0.60, 0.90 over each group, and 0.79 over all configurations.

We also investigated the relationship between the sample size and properties of the e-QTL position estimates via bias and precision (Figure 5). The estimation of the *structural* locus position revealed that both the bias and standard deviation were large when the loci had small effects (0.10). However, estimates of the *regulatory* locus position were relatively unbiased and obtained with higher precision than the *structural* locus position estimates. Bias and poor precision can be addressed and improved upon statistically by using a larger number of individuals in a segregating population.

Based upon the *JZmapqtl* analyses, the power results from this simulation study demonstrate that a multiple gene/trait QTL approach has great potential, but is limited by the statistical model that is implemented for traditional QTL analysis. Through further investigations we have found that increased replicates provide limited additional improvement in statistical power. In order to improve upon the limitations of using standard multiple trait QTL, as applied to microarray data, we propose (elsewhere) a novel statistical model that accounts for the genetic and genomic components of this analysis. The statistical power for all configurations can be improved greatly if replicate gene features are employed, and if the technological variation of the microarray experiments is acknowledged in the decomposition of expression level polymorphisms.

## 5. DISCUSSION

Previous studies that have addressed genetical genomics (Jin et al., 2001; Brem et al., 2002; Steinmetz et al., 2002; Wayne and McIntyre, 2002; Borevitz et al., 2003; Schadt et al., 2003) have used gene expression data based on only one parental line, or one treatment condition related to a complex trait. The e-QTL investigation proposed here exploits gene expression levels quantified under different treatment conditions to identify determinants of ELPs. Our investigation and suggested approach relies on an application of multiple trait interval mapping to detect e-QTLs by providing a way to account for possible interactions between regulatory loci and the treatment. Compared to the standard interval mapping methods (Schadt et al., 2003) that have been used to reveal regulatory and structural regions, our approach provides a statistically more powerful avenue to detect and locate e-QTL due to the benefit of treating the correlated gene expression measurements within each gene as multiple traits and consideration of the statistical interactions involving e-QTL.

Based upon the *JZmapqtl* analyses, the results from this simulation study demonstrated that a multiple trait QTL approach has great potential, but is limited by the statistical model that is implemented for traditional QTL analysis. Through further investigation we found that increased replication provided limited additional improvement in statistical power. The statistical power for all configurations can also be improved greatly if replicate gene features on an array are employed and if the technological variation of the microarray experiments is acknowledged in the partitioning of sources of variation in ELP data. In order to improve upon the limitations of using standard multiple trait QTL methodology on microarray data, we are developing a novel statistical model that accounts for the genetic and genomic components of this analysis.

While both genetics and genomics have provided information on components of regulatory systems, recent progress in technology and analytic methodology is providing ways to analyze biological complexity at systems biological level (Chong and Ray, 2002; Kitano, 2002; Jansen, 2003). The molecular dissection of complex traits is one of the next greatest challenges to be addressed. A comprehensive understanding of regulatory networks determining complex traits is an integral part of this approach. It will be interesting to determine what proportion of ELPs are controlled by regulatory loci and how many are determined at structural loci. We may find that the QTL, which are known to come and go between repeated experiments, environments, and conditions are actually more involved in gene regulation, and that these QTLs may tend to be regulatory e-QTLs rather than structural e-QTLs.

ACKNOWLEDGEMENTS. The work is supported by a NSF 2010 Project grant (115109-MCB) to DAS, RWD, and RWM, and USDA grant (00-52100-9615) to RWD. RWD thanks Dr. Jim Birchler for discussions that have directly benefited this work.

# 6. REFERENCES

Basten, C.J., Weir, B.S., and Zeng, Z.-B., 1994, Zmap—A QTL Cartographer, In Smith *et al.*, C., editor, *In Proceedings of the 5th World Congress on Genetics Applied to Livestock Production: Computing Strategies and Software*, volume 22, pages 65–66, Guelph, Ontario, Canada, 5th World Congress on Genetics Applied to Livestock Production, Organizing Committee.

Basten, C.J., Weir, B.S., and Zeng, Z.-B., 2002, QTL *Cartographer*, Department of Statistics, North Carolina University, Raleigh, NC, Version 1.16.

Black, M.A., and Doerge, R.W., 2002, Calculation of the minimum number of replicate spots required for detection of significant gene expression fold change in microarray experiments, *Bioinformatics*, 18:1609–1616.

Borevitz, J.O., Liang, D., Plouffe, D., Chang, H., Zhu, T., Weigel, D., Berry, C.C., Winzeler, E., and Chory, J., 2003, Large-scale identification of single-feature polymorphisms in complex genomes, *Genome Research*, 13:513–523.

Brem, R.B., Yvert, G., Clinton, R., and Kruglyak, L., 2002, Genetic dissection of transcriptional regulation in budding yeast, *Science* 296:752–755.

Broman, K.W., and Speed, T.P., 2002, A model selection approach for the identification of quantitative trait loci in experimental crosses, *J. R. Statist. Soc. B*, 64:641–656.

Chong, L., and Ray, L.B., 2002, Whole-istic biology, *Science*, 295:1661.

Churchill, G.A., 2002, Fundamentals of experimental design for cdna microarrays, *Nature Genetics Supplement*, 32:490–496.

Churchill, G.A., and Doerge, R.W., 1994, Empirical threshold values for quantitative trait mapping, *Genetics*, 138:963–971.

Cowles, C.R., Hirschhorn, J.N., Altshuler, D., and Lander, E.S., 2002, Detection of regulatory variation in mouse genes, *Nature Genetics*, 32:432–437.

Doerge, R.W., 2002, Mapping and analysis of quantitative trait loci in experimental populations, *Nature Reviews Genetics*, 3:43–52.

Duggan, D.J., Bittner, M., Chen, Y., Meltzer, P., and Trent, J.M., 1999, Expression profiling using cDNA microarrays, *Nature Genetics Suppliment*, 21:10–14.

Eisen, M.B., Spellman, P.T., Brown, P.O., and Botstein, D., 1998, Cluster analysis and display of genome-wide expression patterns, *Proc. Natl. Acad. Sci. USA*, 95:14863–14868.

Holter, N.S., Mitra, Maritan, A., Cieplak, M. Banavar, J.R., and Fedoroff, N.V., 2000, Fundamental patterns underlying gene expression profiles: simplicity from complexity, *Proc. Natl. Acad. Sci. USA*, 97:8409–8414.

Hughes, T., Marton, M., Jones, A., Roberts, C., Stoughton, R., Armour, C., Bennett, H., Dai, H., He, Y., Kidd, M., King, A., Meyer, M., Slade, D., Lum, P., Stepaniants, S., Shoemaker, D., Gachotte, D., Chakraburtty, K., Simon, J., Bard, M., and Friend, S., 2000, Functional discovery via a compendium of expression profile, *Cell*, 102:109–126.

Jang, W., Chen, H.C., Sicotte, H., and Schuler, G.D., 1999, Making effective use of human genomic sequence data, *Trends in Genetics*, 15:284–286.

Jansen, R.C., 2003, Studying complex biological systems using multifactorial perturbation, *Nature Reviews Genetics*, 4:145–151.

Jansen, R.C., and Nap, J.P., 2001, Genetical genomics: the added value from segregation, *Trends in Genetics*, 17:388–391.

Jiang, C., and Zeng, Z.-B., 1995, Multiple trait analysis of genetic mapping for quantitative trait loci, *Genetics*, 140:1111–1127.

Jin, W., Riley, R.M., Wolfinger, R.D., White, K.P., Passador-Gurgel, G., and Gibson, G., 2001, The contributions of sex, genotype and age to transcriptional variance in *Drosophila melanogaster*, *Nature Genetics*, 29:389–395.

Kerr, M.K., and Churchill, G.A., 2001a, Bootstrapping cluster analysis: assessing the reliability of conclusions from microarray experiments, *Proc. Natl. Acad. Sci. USA*, 98:8961–8965.

Kerr, M.K., and Churchill, G.A., 2001b, Experimental design for gene expression microarrays, *Biostatistics*, 2:183–201.

Kerr, M.K., and Churchill, G.A., 2001c, Statistical design and the analysis of gene expression microarray data, *Genet. Res., Camb.*, 77:123–128.

Kerr, M.K., Martin, M., and Churchill, G.A., 2000, Analysis of variance for gene expression microarray data, *Journal of Computational Biology*, 7:819–837.

Kitano, H., 2002, Systems biology: a brief overview, *Science*, 295:1662–1664.

Kodadek, T., 2001, Protein microarrays: prospects and problems, *Chem. Biol.*, 8:105–115.

Lander, E.S., and Botstein, D., 1989, Mapping mendelian factors underlying quantitative traits using rflp linkage maps, *Genetics*, 121:185–199.

Lincoln, S.E., Daly, M.J., and Lander, E.S., 1992, *Mapping Genes Controlling Quantative Traits with MAPMAKER/QTL*, Whitehead Institute for Biomedical Research, Cambridge, MA 02142 USA, two edition, Version 1.1.

Lipshutz, R.J., Fodor, S.P.A., Gingeras, T.R., and Lockhart, D.J., 1999, High density synthetic oligonucleotide arrays, *Nature Genetics Suppliment*, 21:20–24.

MacBeath, G., 2002, Protein microarrays and proteomics, *Nature Genetics Suppliment*, 32:526–532.

Mackay, T.F., 2001, Quantitative trait loci in D*rosophila, Nature Reviews Genetics*, 2:11–20.

Mauricio, R., 2001, Mapping quantitative trait loci in plants: uses and caveats for evolutionary biology, *Nature Reviews Genetics*, 2:370–381.

Mouse Genome Sequencing Consortium, 2002, Initial sequencing and comparative analysis of the mouse genome, *Nature*, 420:520–562.

Newton, M.A., Kendziorski, C.M., Richmond, C.S., Blattner, F.R., and Tsui, K.W., 2001, On differential variability of expression ratios: Improving statistical inference about gene expression changes from microarray data, *Journal of Computational Biology*, 8:37–52.

Oleksiak, M.F., Churchill, G.A., and Crawford, D.L., 2002, Variation in gene expression within and among natural populations, *Nature Genetics*, 32:261–266.

Rossant, J., and Spence, A., 1998, Chimeras and mosaics in mouse mutant analysis, *Trends in Genetics*, 14:358–363.

Schadt, E.E., Monks, S.A., Drake, T.A., Lusis, A.J., Che, N., Collnayo, V., Ruff, T.G., Milligan, S.B., Lamb, J.R., Cavet, G., Linsley, P.S., Mao, M., Stoughton, R.B., and Friend, S.H., 2003, Genetics of gene expression surveyed in maize, mouse, and man, *Nature*, 422:297–302.

Schena, M., Shalon, D., Heller, R., Brown, A.C.P.O., and Davies, R.W., 1996, Parallel human genome analysis: Microarray-based expression monitoring of 1000 genes, *Proc. Natl. Acad. Sci. USA*, 93:10614–10619.

Searle, S.R., 1971, *Linear Models*, John Wiley and Sons, Inc., New York, NY.

Service, R.F., 1999, Physics: The automated approach to protein structure, *Science*, 285:1345.

Steinmetz, L.M., Sinha, H., Richards, D.R., Spiegelman, J.I., Oefner, P.J., McCusker, J.H., and Davis, R.W., 2002, Dissecting the architecture of a quantitative trait locus in yeast, *Nature*, 416:326–330.

The Arabidopsis Genome Initiative, 2000, Analysis of the genome sequence of the flowering plant A*rabidopsis Thaliana, Nature*, 408:796–815.

Wayne, M.L., and McIntyre, L.M., 2002, Combining mapping and arraying: an approach to candidate gene identification, *Proc. Natl. Acad. Sci. USA*, 99:14903–14906.

Yan, H., Dobbie, Z., Gruber, S.B., Markowitz, S., Romans, K., Giardiello, F.M., Kinzler, K.W., and Vogelstein, B., 2002a, Small changes in expression affect predisposition to tumorigenesis, *Nature Genetics*, 30:25–26.

Yan, H., Yuan, W., Velculescu, V.E., Vogelstein, B., and Kinzler, K.W., 2002b, Allelic variation in human gene expression, *Science*, 297:1143.

Zeng, Z.-B., 1993, Theoretical basis for separation of multiple linked gene effects in mapping quantitative trait loci, *Proc. Natl. Acad. Sci. USA*, 90:10972–10976.

Zeng, Z.-B., 1994, Precision mapping of quantitative trait loci, *Genetics*, 136:1457–1468.

*Chapter 8*

# Charting Contig–Component Relationships within the Triticeae

# Exploiting the Genome

Gerard R. Lazo, Nancy Lui, Frank M. You, David
D. Hummel, Shiaoman Chao, and Olin D. Anderson

## 1. INTRODUCTION

Expressed sequence tags (ESTs) in general reflect the diversity of gene expression in living organisms, and Triticeae ESTs in particular show this diversity in plants. These sequences result from an established path in the laboratory: pieces of single-stranded messenger ribonucleic acid (mRNA) are isolated from plant tissue, converted into double-stranded complementary deoxyribonucleic acid (cDNA), cloned into vector replicons, then transformed into *Escherichia coli* for replication. Deoxyribonucleic acid (DNA) is extracted from these clones and sequenced using high-throughput methods resulting in pools of EST data (Adams, 1992). Such methods mean that a given set of sequences, often called a "library", shares a common origin, i.e., they have the same species, cultivar, tissue, condition, and stress attributes. Their characteristics represent a snapshot of the organism, captured at the point in time when the researcher isolated the mRNA.

**Gerard R. Lazo and Olin D. Anderson**    United States Department of Agriculture, Agricultural Research Service, Western Regional Research Center, Albany, California 94710-1105.    **Nancy Lui, Frank M. You, David D. Hummel, and Shiaoman Chao**    University of California, Davis, California 95616.

*Genome Exploitation: Data Mining the Genome*, edited by J. Perry Gustafson, Randy Shoemaker, and John W. Snape.
Springer Science + Business Media, New York, 2005.

The abundance of EST data has increased dramatically in the past few years. The plant tribe Triticeae includes several closely related crop plants of major economic importance, including wheat, barley and rye (Barkworth et al., 1992, Kellogg, 2002). Only a handful of ESTs from the species *Triticum aestivum*, bread wheat, were available in the year 1998; now the numbers for this and other tribe Triticeae species number over 750,000 (NCBI dbEST, 2003). The EST information is available to the public through contributions submitted to the NCBI Genbank resources (Boguski et al., 1993), and much of the accompanying Triticeae information relating to these specific ESTs is available at the GrainGenes project site (Matthews et al., 2003). These sequences have been applied to Triticeae genomics in a wide variety of ways, including the development of molecular markers, placement on physical and genetic maps, characterization as gene candidates, and used for comparative studies between related species (Akhunov et al., 2003, Sorrells et al., 2003).

To remove redundancy from within an EST data set, sequences are aligned and clustered using various assembly algorithms, some of the more popular being CAP3 (Huang and Madan, 1999), phrap (Green, 2003) and d2_cluster (Burke et al., 1999). In building an assembly, a set of unique gene sets can be assembled into "unigenes", essentially representing a range of genes present in an organism (Pontius et al., 2003, Liang et al., 2000, Quakenbush et al., 2000). The success of an assembly relies on the quality of the sequence data and the various parameters available within the software used to provide the established settings for sequence-by-sequence comparisons.

A software tool has been developed to complement gene discovery efforts by providing an overall visual representation of a sequence assembly, or clustering. The software interacts with a relational database to point to other relevant pieces of information. Rather than having to develop specific questions to query a relational database, the display produces a global perspective of the data set and allows orienting decisions based on attributes of the contributing data variables, allowing the observer to make intuitive research decisions based on the clustering patterns of the data elements. Data points may be selected individually or collectively to follow in-depth information associated with the data points. With this tool it is possible to pose a wide range of queries based on known data point features such as those relating to library origin, metabolic pathway, map position, and the like. Applications of this tool are primarily directed at EST analysis. This software has also been modified to interact with data for microarray expression analysis, and phylogenetic determination of genome sequences.

## 2. APPROACH

For use in the study of EST data a visualization tool was constructed, named Contig Constellation Viewer (CCV), to relate to data variables of cDNA libraries and their contributions toward assembled contigs. To relate to potential expression profiles, the libraries can be sorted based on, for example, species, cultivar, tissue,

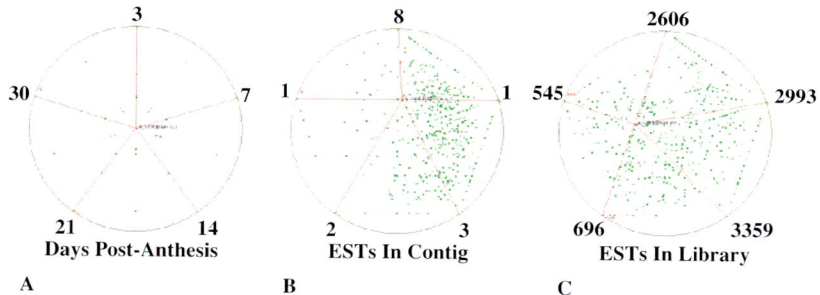

**FIGURE 1.** Differences in CCV displays. The three display algorithms used above are *Equal* (**A**), *Proportional* (**B**), and *Weighted* (**C**). All of the images above represent the same dataset. The numbers shown in (**A**) represent the 3, 7, 14, 21, and 30 days post-anthesis cDNA library points equally spaced around the circle with a single contig highlighted showing connections to the libraries having ESTs in the contig (in this sample all libraries contribute to the contig). The numbers in (**B**) represent the number of ESTs from each library found in Contig 8694. The numbers in (**C**) indicate the number of ESTs sequenced in each library, thus there are 10,199 ESTs represented in the 2,051 contigs shown for each data set representation above.

developmental stage, or stresses. Clustering patterns would reflect those contigs with an abundance of ESTs important for the different sorted criteria. The orientation of contigs displayed is dependent on the sorting order and the algorithm utilized to represent the display. With the sampled assemblies, up to 50,000 contigs were represented at a time with the visualization tool. The patterned layout of contigs in the CCV display would provide an intuitive means to focus on contigs representing interests for the researcher.

## 2.1. Display Settings

Three different display algorithm settings were applied, termed: *equal, proportional*, and *weighted*. Each setting represented how the EST libraries placed around the circumference of the graphical display influenced the spatial placement of contigs within the circle. In the example provided in Figure 1, only five libraries are displayed of the 152 libraries used in the assembly to simplify showing how the differently applied algorithms affect placement of contigs within the display. Each of the three samples represents the same data, but each algorithm applied had benefits for navigating the contigs and associating supplemental information available for each of the contigs. In this example, the five-library subset was also selected to add some insight toward the expression of ESTs derived from cDNA libraries prepared at different stages of kernel development (Tingey et al., 2003).

### 2.1.1. Equal Setting

This setting placed contigs within the display in a non-discriminant manner with respect to the libraries contributing to contigs (Figure 1A). The placement of a contig represented a clear mid-point cross-section between all libraries contributing at least one EST to the assembled contig. If every library represented in the display had at least one EST member in a particular contig, that contig would be placed at the center point of the CCV display as shown in Figure 1A.

### 2.1.2. Proportional Setting

This setting placed contigs in the display with respect to the number of EST members from a given library represented in a contig (Figure 1B). If a single library contributed proportionally 0.5, or one half, of the ESTs clustered into a contig, the contig point would migrate 0.5 of the distance, between the relevant libraries, towards the direction of the represented library. Similarly, the other library influences on point migration would be determined by the proportional representation of the other libraries in the contig.

### 2.1.3. Weighted Setting

This *Weighted* setting is much like the *Proportional* setting in that the number of ESTs contributing to the contig is important. However, to account for contributions from libraries from which few ESTs were sequenced, the contig point migration is adjusted based on fractional representation of the ESTs from a given library (Figure 1C). For instance, a single contig with ESTs derived from two libraries, consisting of one EST from library A and nine ESTs from library B, would be located midpoint if library A had a total of 100 sequences and library B had a total of 1,000; both libraries would be given a weighted value of 0.01, each representing about one percent of the library in the contig.

The above were only three of the settings initially tested. It would be possible to add other sorting algorithms to provide different graphical perspectives.

### 2.2. Discerning Contig-Component Relationships

Because libraries represented in the CCV display can be sorted based on key library components, or attributes, the display can be geared towards uncovering the way contigs associate with these key attributes. Since contigs patterned as simple intersections between libraries may not adequately explain contig-component relationships, other algorithm settings can be used. Following are a few case studies representing a range of applications.

### 2.2.1. Tissue Differentiation

The ordering of the libraries within the CCV display by tissue attributes makes it possible to detect contigs that may be strongly associated with specific tissues. Sorting the libraries by tissue could be useful when a researcher is focusing on contigs associated with tissue-specific expression or when hunting for genes that may carry tissue-specific promoters. As an example, Fusarium head blight is a serious disease threat to the Triticeae agricultural crops, primarily affecting the flowering parts of the plant (Cook, 1981). Attempts at constructing or studying potential resistance mechanisms are focusing on expression in the spike tissues. The sorting of spike and closely related tissues together in the CCV display facilitates analysis of those contigs highly associated with those tissues.

### 2.2.2. Developmental Expression

A single assembly may involve constructing contigs from sequences derived from a wide range of libraries; there are about 152 different cDNA libraries constructed for *T. aestivum*. In some cases, cDNA libraries were constructed to cover a very distinct set of developmental stages, for instance: 3, 7, 14, 21, and 30 days after anthesis for the developing kernel (Tingey et al., 2003). Displaying only a subset of these five libraries allows a visualization of contigs associated with each stage of development, as well as those constitutively expressed (Figure 1). Visualizing the contigs in this fashion allow for stepwise selection of contigs specifically associated between incremental steps. Likewise, such a collection could also be compared to libraries for which several time points may have been pooled. The ability to study library subsets, and depending on the range of stages from which libraries have been constructed leads to the opportunities to study developmental traits relating to flowering, nutrition, resistance/susceptibility, among other potential quantitative traits.

### 2.2.3. Treatment Differentiation

Many cDNA libraries have been constructed primarily to distinguish differences in expression between different conditions, and a variety of methods have been developed to exploit these differences (e.g. differential display, microarrays, subtractive hybridization). The CCV display can be adjusted to pool different treatments and match them up against controls. There are publicly available libraries that have been constructed against a range of pathogens and environmental stresses allowing the possibility to categorize genes associated with different conditional states of the plant.

### 2.2.4. Germplasm Differentiation

There may be situations where a comparison between different germplasm may explain the differences in gene expression. It may be possible to identify genes responsible for quality phenotypes, disease resistance, or differential expression due to nutrition or stress conditions. In cases where phenotypes are multigenic, a family of genes associated with a quality trait might be determined. The CCV

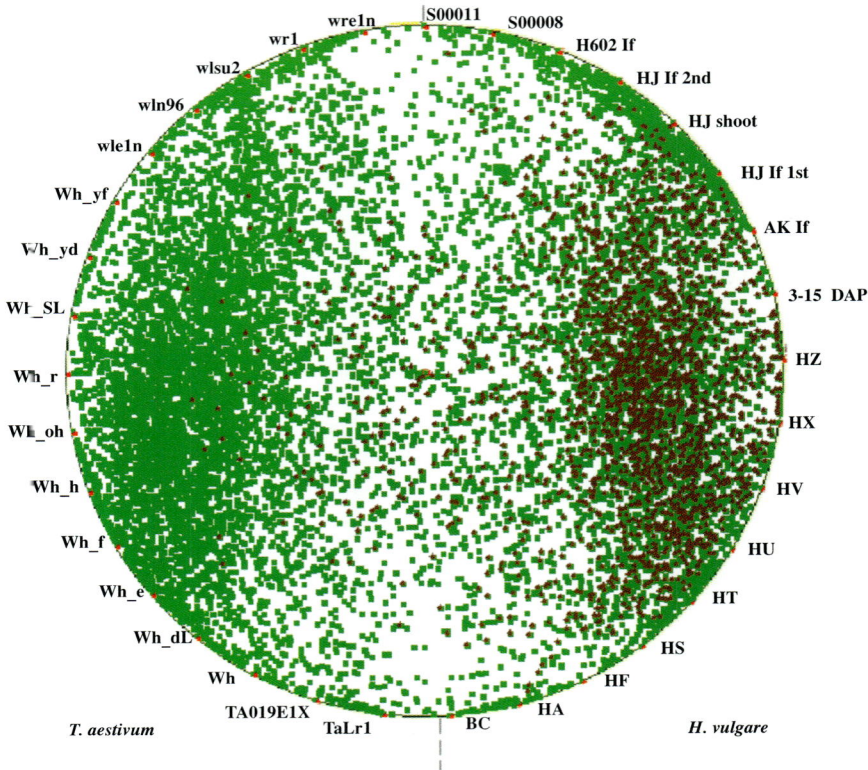

**FIGURE 2.** A co-assembly of wheat and barley ESTs. Phrap assembly was performed on 776,000 ESTs from a total of 264 cDNA libraries from *Triticum aestivum* and *Hordeum vulgare* species. Shown are the 34 largest libraries, 17 from each species, with 19,493 contigs shown representing 280,226 ESTs. Each dot represents a contig, and libraries contributing to that contig using a weighted algorithm represent its position. Highlighted is an *H. vulgare* library (**HZ**) showing all the contigs, which have ESTs from that library present in the highlighted contig.

display allows for the sorting of assembled genes that are shared or differentiated by germplasm.

As contigs can be assembled to differentiate germplasm, such as by cultivar characteristics; the same display can also be used to differentiate germplasm by species. Depending on the stringency of the assembly it may be able to distinguish between genes shared, or highly similar between germplasm and species. In the example provided, an assembly was set up to compare the species *T. aestivum* and *Hordeum vulgare* (Figure 2). The general observation was that the species were quite distinct, but there were many instances where the contigs formed contained and shared ESTs derived from both species.

## 2.3. Other Applications

### 2.3.1. Microarray Comparison

Microarray technology is rapidly becoming a primary means for measuring gene expression due to capabilities to use high throughput means to create the arrays and the ability to screen thousands of genes at one time (Fellenberg et al., 2001, Li, 2001). The probes used to screen against microarrays are similar to the building of cDNA libraries in that the probe material is derived from the isolation of mRNA under a specified set of conditions. Microarray studies are just now under development for study within the Triticeae species; however, to facilitate the study of microarray expression analysis, the CCV display tool was initially set up to read currently available data sets to perform mock microarray analysis without the costly production of microarrays or building of RNA probes. For instance, microarray experiments for the model organism *Arabidopsis thaliana* was used for study and compared against a *T. aestivum* contig assembly (Figure 3). In this case, the Triticeae contigs were matched to microarray probe sequences using different threshold cutoff values. Also added, was a false-coloring overlay to give a appearance similar to that associated with microarray analysis software. Though the data is derived from another species, this is a feature may assist in pointing to previously documented genes for which expression is somewhat understood and may point to new or unrealized gene relationships based on expression profiles and clustering.

### 2.3.2. Assembly Comparison

For phylogenetic studies, a series of stepwise calculations are performed to build independent data assembly sets; migration of clusters, or changes in the cluster members can be analyzed to determine sequence, or phylogenetically-dependent associations. Given another assembly, it may be possible to distinguish sequences that are genome specific from within a polyploid environment. For instance, *T. aestivum* is a hexaploid species having genome content from the A, B. and D genomes. Using a combination of hexaploid, tetraploid, and diploid species the display set would be sorted by species to observe genome-specific clustering. By comparing different assembly methods, patterned changes in clustering may lead to develop observations on genome evolution as determined by cluster formation.

### 2.3.3. Other Comparisons

Through the assembly method, several representatives of a gene class may be present in the display which may represent duplication of related sequences; this is especially so in a polyploid organism. A simple query will show the placement of the related sequences as distributed with the sorting criterion. A follow-up analysis

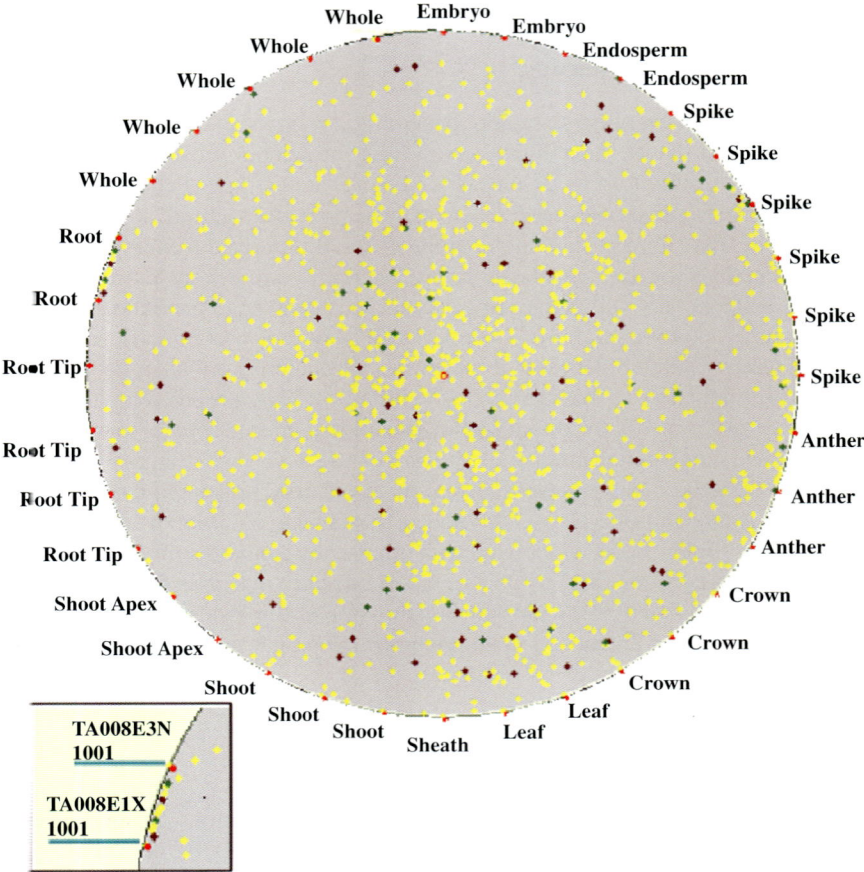

**FIGURE 3.** Microarray Overlay Using CCV Display. Shown are data from 35 wheat cDNA libraries which generated 11,758 contigs from 57,885 ESTs. Using false-coloring derived from a microarray experiment on *Arabidopsis thaliana*, probes used in that experiment were best matched to wheat contigs; the contig color represents co-expressed signals (yellow) compared to both the experiment (green) and the corresponding control (red). The experiment shown involved inoculation of *A. thaliana* with *Heterodera schachtii* nematode. Interestingly, three of the four differentially expressed signals associated with root tissue corresponded to contigs for which ESTs were mapped to chromosome 2B in wheat, a gene location associated with *H. avenae* interactions. Data was derived from the Nottingham Arabidopsis Stock Centre.

of the contigs formed may point to the cause of contig divergence, possibly due to sequence evolution and formation of homoeologous loci due to duplications or rearrangements. Or it may simply be the lack of closing a gap between 5′ and 3′ sequences. However, by a thorough study of contig placements within the CCV display, it may be possible to develop theories of gene adaptation, which can be associated with certain tissues, or stages of development.

With respect to the sorting criterion, contig-component relationships may be displayed to relate clustering, or scattering. For instance, one may be interested in how specific pathway-associated sequences are related to the libraries displayed. By loading a list of identified pathway-classified contigs, only those contigs can be highlighted and can relate candidate function to map position, metabolic pathway, or other general interest queries. In some cases this is useful for assessing the quality of the library. If the library was from a subtracted or normalized library, the treatment can be easily compared to one that was not treated. This is sometimes useful for determining if the library is yielding additional unique sequences, or if the general background signals are being minimized.

## 3. SOFTWARE

The primary software interface is a Java applet (Java, 2003), which interacts with information housed in a relational database. The relational database, mySQL in this case (MySQL, 2003), houses the relevant information for relating contigs to components of associated libraries. Supplemental information for contigs and their sequences are also served through the relational database. For convenience, the Java applet was served through an Internet browser interface as a client application. On the server side, a Java servlet queried through a web server, such as Apache (Apache, 2003), to retrieve data from the relational database. For each assembly set up for study, a protocol was established for uploading the data sets.

## 4. DISCUSSION

Much of the genome sequence data needed is often housed in relational databases and can be readily retrieved, but in order to query the database it is important to know what questions need to be asked. From an object or relational database perspective, the data is viewed very sparingly, derived solely from the questions poised directly to the database. The CCV application presented here attempts a global visualization of the database in a spatial array that can be further queried in various ways making use of its links to a relational database. Curators often conceive of the most probable queries, but many do not address the full breadth of the queries that the researcher may have in mind, or be in search of. The presented CCV tool attempts to provide an overview of the available data and by ranking the library attributes by criteria of interest to the researcher, and provides an intuitive interface that will help the researcher focus on those sequences or candidates that may be most applicable to the research needs. In consideration of the enormous amount of EST information available, it appears to be a daunting task to relate to sequences collectively by relating which attributes are most important and are those sought by the researcher. The setup of the CCV interface attempts to give the researcher an overview perspective of all available data in an assembly study and relate it to a collection of attributes believed to be useful in sorting the data. The graphical interface simplifies the ability to create queries and provides an interface to make general observations and develop new queries. The interface uses one assembly and uses contigs for studying sequences based on their derived origins.

Considering that the over 780,000 sequences representing the Triticeae are derived from only 256 cDNA libraries, it is still unclear to what extent the expressed portion of the genome is represented. It is possible to gauge these numbers from model organisms with sequenced genomes, but even in these cases, the numbers are still unclear. In surveying the expression association patterns with some of the Triticeae assemblies, it appears clear that ESTs fall short of representing the full diversity of expressed genes possible. For example, sequences from a cDNA library derived from callus tissue in barley were found to have a high number of EST sequences in contigs that were apparently callus-specific. Many of the sequences here were apparently unique from those expressed in other tissues. This suggests that an abundance of sequences derived from callus have not been detected under classic mRNA conditional states of isolation. It is also suggested here that callus-derived tissues may relate to a state where inhibitions of mRNA transcription are released, allowing a diversity of sequences to be expressed and detected, except out of context with respect to tissue, development, and the like. Sequences of this sort may simply define the state of callus expression.

From the above simple observation, there still appears to be value in sequencing additional cDNA libraries under a wider range of expression profiles. This would add additional value to the cDNA libraries by enabling the tracking of added attributes, or variables, such as species, germplasm, tissue, developmental stage, and stresses. Establishing new unstudied conditions for the production of cDNA libraries would be useful for coaxing a genome to express sequences important to different states of being, including relationships to developmental stage, stresses, or tissue types.

The computational design presented here was an attempt to condense large datasets into a manageable and discernable environment using visualization methods. The approach used here was different in that contigs could be viewed based on their sequence construction from assembly algorithm programs, allowing each contig to be related in context to other assembly contigs based on their sources and attributes. The end result is a global display of the assembly experiment, with the ability to study more in detail the clusters that are formed by connecting to an accompanying relational database, which contains supplemental information for the experiment. If needed, a subset of the assembly could be displayed to uncover noted interactions. The global visualization in many cases simplified access to the data that normally would require several directed queries to uncover the same information. In other cases the visualization provided a display that would prompt the user to intuitively query for data, which would not be obvious starting from a "command-line" type query. In some sense the visualization of the data analysis duplicates the use of housed database information, but provides a different perspective of the data set, and provides an interface by which to view and query the data. There is a need for database tools that go beyond a simple listing of query results. Tools need to be designed to allow the user a chance to interact and explore in a quest for gene discovery.

ACKNOWLEDGEMENTS.  The work shown here is part of the United States Department of Agriculture, Agricultural Research Service Current Research Information System (CRIS no. 0404546), and partial funding was provided by the National Science Foundation Award (NSF no. 9975989).

## 5. REFERENCES

Adams, M.D., Dubnick, M., Kerlavage, A.R., Moreno, R., Kelley, J.M., Utterback, T.R., Nagle, J.W., Fields, C., and Venter, J.C., 1992, Sequence identification of 2375 human brain genes, *Nature* 355:632–634.

Akunov, E.D., Goodyear, J.A., et al., 2003, The organization and rate of evolution of the wheat transcriptome are correlated with recombination rates along chromosome arms, *Genome Res.* 13:753–763.

Apache HTTP Server. Apache Software Foundation, Forest Hill, MD, USA (March, 2003); http://www.apache.org.

Barkworth, M.E., 1992, Taxonomy of the Triticeae: a historical perspective, *Hereditas* 116:1–14.

Boguski, M.S., Lowe, T.M., and Tolstoshev, C.M., 1993, dbEST—database for "expressed sequence tags." *Nat. Genet.* 4:332–333.

Burke, J., Davison, D., and Hide, W., 1999, d2_cluster: a validated method for clustering EST and full-length cDNA sequences, *Genome Res.* 9:1135–1142.

Cook, R.J., 1981, Fusarium diseases of wheat and other small grains in North America. In: *Fusarium: Diseases, Biology and Taxonomy*, Nelson, P.E., Tousson, T.A., Cook, R.J., eds, Pennsylvania State Univ. Press. University Park, pp: 39–52.

Fellenberg, K., Hauser, N.C., Brors, B., Neutzner, A., Hoheisel, J.D., and Vingron, M., 2001, Correspondence analysis applied to microarray data, *Proc. Natl. Acad. Sci.* USA 98:10781.

Green, P., University of Washington, Seattle, WA. Phrap program (March, 2003); http://www.phrap.org.

Huang X., and Madan, A., 1999, CAP3: A DNA sequence assembly program, *Genome Res.* 9:868–877.

Java 2 Platform. Sun Microsystems, Palo Alto, CA (March, 2003); http://java.sun.com.

Kellogg, E., 2001, Evolutionary History of the Grasses, *Plant Physiol.* 125:1198–1205.

Li, K.-C., 2001, Genome-wide coexpression dynamics: theory and application, *Proc. Natl. Acad. Sci.* USA 99:16875–16880.

Liang, F., Holt, I., Pertea, G., Karamycheva, S., Salzberg, S.L., and Quackenbush, J., 2000, An optimized protocol for analysis of EST sequences, *Nucleic Acids Res.* 28:3657–65.

Matthews, D.E., Carollo, V.L., Lazo, G.R., and Anderson, O.D., 2003, GrainGenes, the genome database for small-grain crops, *Nucleic Acids Res.* 31:183–186.

MySQL Database Server, MySQL, Inc., Seattle, WA, USA (March, 2003); http://www.mysql.com.

National Center for Biotechnology Information, 2003, Bethesda, Maryland (NCBI dbEST, March, 2003); http://www.ncbi.nlm.nih.gov/dbEST.

Nottingham Arabidopsis Stock Centre, Affymetrix Service, University of Nottingham, University Park (Nottingham, UK, March, 2003); http://arabidopsis.info.

Quackenbush, J., Liang, F., Holt, I., Pertea, G., and Upton, J., 2000, The TIGR gene indices: reconstruction and representation of expressed gene sequences, *Nucleic Acids Res.* 28:141–145.

Radchuk, V., Zhang, H., Weschke, W., Potokina, E., and Wobus, U., 2002, EST library HZ and others deposited at NCBI dbEST (March, 2003); http://www.ncbi.nlm.nih.gov/dbEST.

Sorrells, M.E., La Rota, C.M., et al., 2003, Comparative DNA Sequence Analysis of Wheat and Rice Genomes, *Genome Res.* (in press).

Tirgey, S.V., Powell, W., Wolters, P., Dolan, M., Hainey, C., Yuan, Z., Miao, G., Caraher, N., and Hanafey, M.K., 2003, EST libraries wdk1c, wdk2c, wdk3c, wdk4c, wdk4c, deposited at NCBI dbEST (March, 2003); http://www.ncbi.nlm.nih.gov/dbEST.

Urwin, P.E., Lilley, C.J., McPherson, M.J., and Atkinson, H.J., 1997, Resistance to both cyst and root-knot nematodes conferred by transgenic Arabidopsis expressing a modified plant cystatin, *Plant J.* 12:455–461.

*Chapter 9*

# Protein Family Classification with Discriminant Function Analysis

Etsuko N. Moriyama and Junhyong Kim

## 1. INTRODUCTION

Rapid progress in multiple genome projects continues to feed databases in the world a large volume of sequence data. In this "post-genomic" era, more efficient and reliable sequence annotation, especially functional annotation of protein sequences, is crucial. Although experimental confirmation is ultimately required, computational annotation of protein sequences has been routinely done, and it is incorporated into major protein databases (*e.g.*, SWISS-PROT: http://www.expasy. org/sprot/, PIR-PSD: http://pir.georgetown.edu/pirwww/search/textpsd.shtml). Due to a rapidly growing number of new sequences, increasingly more database entries contain only computational annotations.

In this paper, we first discuss the disadvantage commonly found in various existing protein classification methods. Next we introduce a set of new methods that can classify protein family sharing very weak similarity. Finally, we describe an algorithm that combines strengths from various protein classification methods to obtain an optimum power for protein classifications.

**Etsuko N. Moriyama**    School of Biological Sciences and Plant Science Initiative, University of Nebraska, Lincoln, Nebraska, 68588-0660.    **Junhyong Kim**    Department of Biology, University of Pennsylvania, Philadelphia, Pennsylvania, 19104-6018.

*Genome Exploitation: Data Mining the Genome*, edited by J. Perry Gustafson, Randy Shoemaker, and John W. Snape.
Springer Science + Business Media, New York, 2005.

Table 1
Protein Classification Tools.

| Tool | Description | Reference |
|---|---|---|
| BLAST | Local sequence similarity search tools (blastn, blastp, *etc.*) | Altschule *et al.* (1990) |
| PRINTS/SPRINT | Protein fingerprint database (searched by FingerPRINTScan) | Attwood *et al.* (2002) |
| PROSITE | Database for biologically significant sites, patterns and profiles | Falquet *et al.* (2002) |
| Pfam | Multiple alignment and profile HMM database (searched by HMMER) | Bateman *et al.* (2002) |
| PSI-BLAST | Position specific iterative BLAST using position specific scoring matrix | Altschul *et al.* (1997) |
| SMART | Domain architecture research tool (profile HMM database) | Letunic *et al.* (2002) |

## 1.1. Protein Classification Methods

In order to improve the power of computational annotations, various methods have been developed. Computational annotation (or classification) methods rely on finding similarity between a query (new protein) sequence and protein sequences in databases with known (preferably experimentally confirmed) functions. The most popularly used method is the Basic Local Alignment Search Tool (BLAST) by Altschule *et al.* (1990). It searches databases for sequences with local similarity to the query. When more distant similarity is sought, pattern or profile, rather than the sequence itself, is used for the database search. Table 1 lists some methods frequently used for protein annotation and protein family classification.

Except BLAST (including PSI-BLAST), all of the search tools listed above have their own pattern, profile, or motif database. These patterns/profiles are generated from alignments of known protein sequences. Since functionally more important regions (*e.g.*, catalytic domains, binding-domains) are considered to be under stronger selective constraints, multiple alignments from proteins with known functions are expected to contain conserved regions related to those functions. When distantly related sequences are compared, only functionally critical sites, rather than a large region, might be conserved. Furthermore, some amino acids may be substituted with others (usually with other biochemically similar amino acids) as long as the protein function is maintained. Pattern and profile search methods allow such flexibility and they are more sensitive to weakly conserved sequences than simple similarity search methods. Even when regular BLAST search fails to identify any significantly similar sequence to the query from the database, pattern/profile search methods frequently can detect a signature pattern/profile related to a known function.

## 1.2. Pros and Cons for the Current Protein Classification Tools

Due to the differences in their underlying techniques and also in their focuses (*e.g.*, family coverage), each method (and database) has different strengths

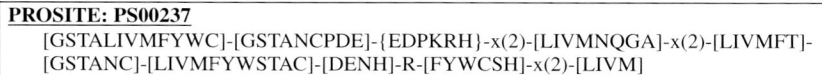

PROSITE: PS00237
[GSTALIVMFYWC]-[GSTANCPDE]-{EDPKRH}-x(2)-[LIVMNQGA]-x(2)-[LIVMFT]-
[GSTANC]-[LIVMFYWSTAC]-[DENH]-R-[FYWCSH]-x(2)-[LIVM]

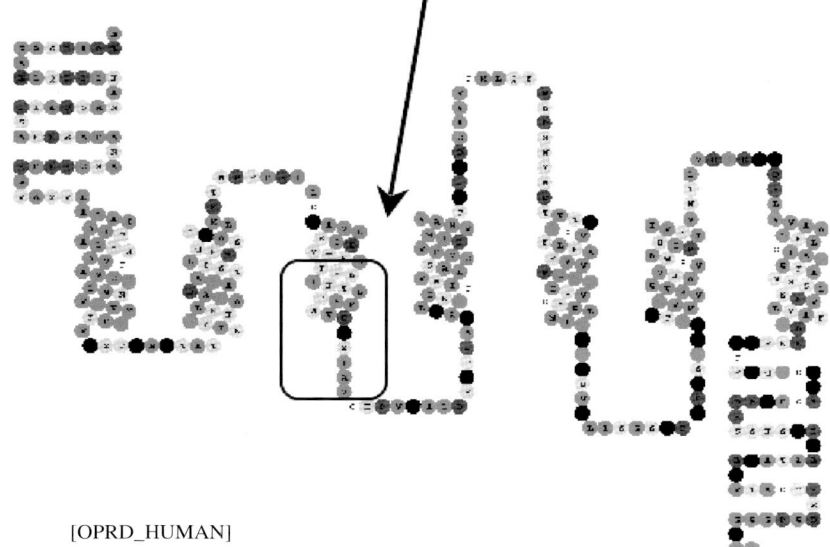

[OPRD_HUMAN]

**FIGURE 1.** An example entry from the PROSITE pattern database. This is one of the pattern entries for G-protein coupled receptors (PS00237). A G-protein couple receptor sequence (a human opioid receptor entry from SWISS-PROT: OPRD_HUMAN) is shown under the PROSITE pattern entry. Each dot represents one amino acid. Seven cylinders in the middle indicate predicted seven transmembrane regions. The circled area on the sequence corresponds to the PROSITE regular expression pattern entry spanning 17 amino acids.

and weaknesses. In order to take maximum advantage from these various information sources, usually it is necessary to conduct multiple pattern/profile searches. Integrated databases, *e.g.*, InterPro (http://www.ebi.ac.uk/interpro/) and MetaFam (http://metafam.ahc.umn.edu/), were developed to facilitate such tedious procedures.

One of the problems inherited in all of these pattern/profile search methods and databases is that their patterns are in general derived from relatively short regions. It is particularly the case in the PROSITE patterns. PROSITE patterns are expressed in regular expressions as shown in Figure 1. If the query is only a partial sequence (*e.g.*, EST), and if it does not contain the region where the pattern was derived, this method fails to identify the query correctly. The regular expressions also allow only limited flexibility.

PRINTS uses also very short conserved motifs. But it tries to overcome this problem by identifying multiple motifs covering a larger region than a single motif. Figure 2 above shows one example PRINTS entry that includes seven fingerprint motifs.

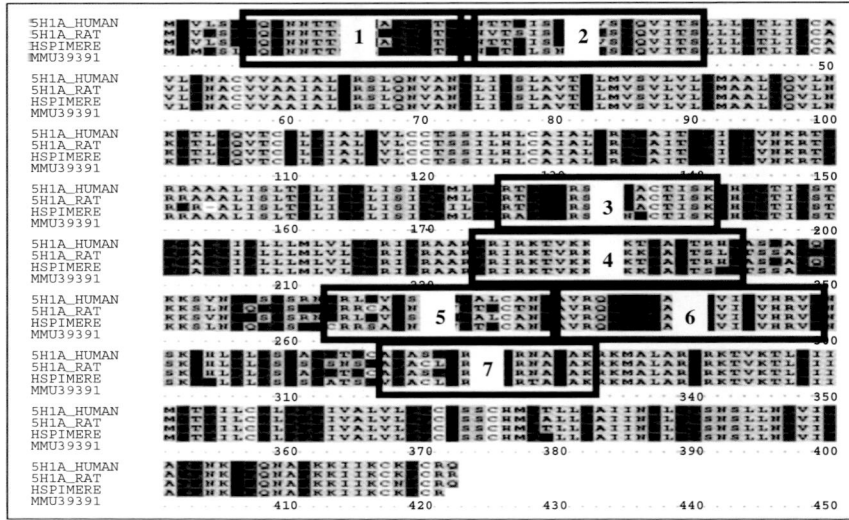

**FIGURE 2.** An example protein fingerprint from the PRINTS database. This is "5-hydroxytryptamine 1A receptor signature" (5HT1ARECEPTR). This PRINTS entry includes seven motifs or "fingerprints." The alignment below (including four 5-hydroxytryptamine 1A receptor sequences) shows the locations of these seven fingerprints (the boxes 1–7).

Profiles (used in PROSITE/Profile and PSI-BLAST) express the flexibility in amino acid substitutions at each position in a series of scoring matrices, "Position Specific Scoring Matrix" (PSSM). The profile hidden Markov model (profile HMM) is a probabilistic model of sequences and used in Pfam and SMART. Profiles and profile HMMs cover the entire region of alignments, usually much longer than regions covered by PROSITE/pattern or PRINTS. Figure 3 below shows an example profile HMM from a Pfam entry.

An inherent problem in these methods is that they rely on multiple alignments for generating the patterns and profiles. However, generating multiple alignments themselves becomes problematic when extremely distant sequences are involved. Furthermore, diagnostic patterns and profiles cannot be easily identified from a multiple alignment that includes extremely diverged sequences.

Yet another problem shared by existing classification methods is that the patterns, motifs, and profiles need to be identified from already known protein sequences. Since subsequently found proteins are classified based on these patterns/profiles, possible initial sampling bias will be reinforced.

## 1.3. G-protein Coupled Receptor Super Family

A good example representing such extreme diversity is the G-protein coupled receptor (GPCR) super family. Many medically and pharmacologically important

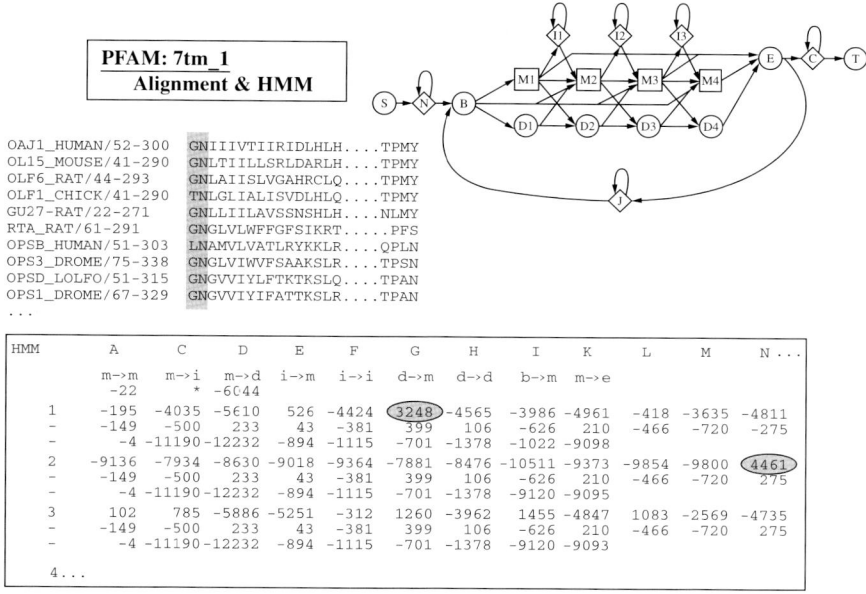

```
PFAM: 7tm_1
Alignment & HMM

OAJ1_HUMAN/52-300   GNIIIVTIIRIDLHLH....TPMY
OL15_MOUSE/41-290   GNLTIILLSRLDARLH....TPMY
OLF6_RAT/44-293     GNLAIISLVGAHRCLQ....TPMY
OLF1_CHICK/41-290   TNLGLIALISVDLHLQ....TPMY
GU27-RAT/22-271     GNLLIILAVSSNSHLH....NLMY
RTA_RAT/61-291      GNGLVLWFFGFSIKRT.....PFS
OPSB_HUMAN/51-303   LNAMVLVATLRYKKLR....QPLN
OPS3_DROME/75-338   GNGLVIWVFSAAKSLR....TPSN
OPSD_LOLFO/51-315   GNGVVIYLFTKTKSLQ....TPAN
OPS1_DROME/67-329   GNGVVIYIFATTKSLR....TPAN
...
```

| HMM | A | C | D | E | F | G | H | I | K | L | M | N ... |
|---|---|---|---|---|---|---|---|---|---|---|---|---|
| | m->m | m->i | m->d | i->m | i->i | d->m | d->d | b->m | m->e | | | |
| | -22 | * | -6044 | | | | | | | | | |
| 1 | -195 | -4035 | -5610 | 526 | -4424 | 3248 | -4565 | -3986 | -4961 | -418 | -3635 | -4811 |
| - | -149 | -500 | 233 | 43 | -381 | 399 | 106 | -626 | 210 | -466 | -720 | -275 |
| - | -4 | -11190 | -12232 | -894 | -1115 | -701 | -1378 | -1022 | -9098 | | | |
| 2 | -9136 | -7934 | -8630 | -9018 | -9364 | -7881 | -8476 | -10511 | -9373 | -9854 | -9800 | 4461 |
| - | -149 | -500 | 233 | 43 | -381 | 399 | 106 | -626 | 210 | -466 | -720 | 275 |
| - | -4 | -11190 | -12232 | -894 | -1115 | -701 | -1378 | -9120 | -9095 | | | |
| 3 | 102 | 785 | -5886 | -5251 | -312 | 1260 | -3962 | 1455 | -4847 | 1083 | -2569 | -4735 |
| - | -149 | -500 | 233 | 43 | -381 | 399 | 106 | -626 | 210 | -466 | -720 | 275 |
| - | -4 | -11190 | -12232 | -894 | -1115 | -701 | -1378 | -9120 | -9093 | | | |
| 4 ... | | | | | | | | | | | | |

**FIGURE 3.** An example entry from Pfam. A part of a G-protein coupled receptor entry (7tm_1) is shown in the box. Two circled numbers correspond to the "emission probability" for a glycine at the first amino acid position and an alanine at the second position, respectively. Note that the amino acids "G" and "N" are the majority at the first and second positions in the alignment above, respectively, and the "emission probabilities" for these two amino acids are the largest at each site. The diagram above is the transition structure of the HMM model used in HMMER (the HMM program package used in Pfam).

proteins are included in this family: *e.g.*, acetylcholine receptors, dopamine receptors, and opioid receptors. Therefore, classifying this protein family and finding new members of this family is one of the most important topics in medical genomics.

The GPCR protein family is one of the most diverse protein families. The family is classified into five major classes (A-E) as well as other minor classes and putative and "orphan" groups. The members of this family share one structural feature, seven-transmembrane regions as shown in Figure 4. Beyond this structural similarity, the members, especially those in different classes share very low sequence similarity. The seven transmembrane regions contribute to the low sequence similarity because many of the hydrophobic amino acids within the region are interchangeable as long as they are hydrophobic and do not disrupt the structural conformation. On the other hand, the loop regions between the transmembrane regions can be varied in length. Therefore, transmembrane and loop regions contribute to the low sequence similarity in different ways. The low sequence similarity and heterogeneity created by repeated transmembrane and loop regions creates the most difficult situation in generating multiple alignments, and no reliable multiple alignment can be generated from the entire super family. The GPCR protein family presents one of the most challenging properties for protein classification

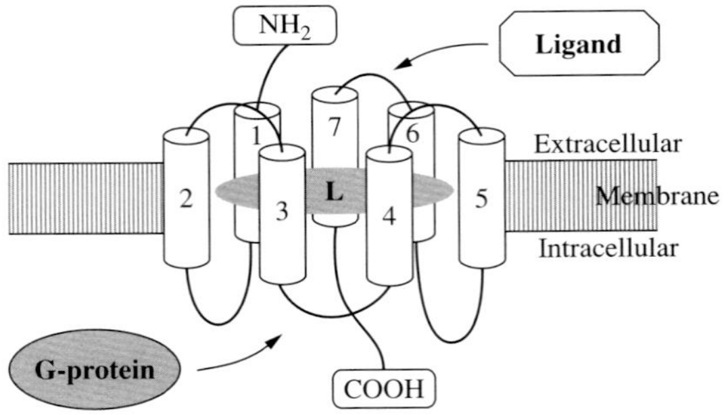

**G-protein Coupled Receptor**

**FIGURE 4.** A model of G-protein coupled receptors. The seven transmembrane regions are shown in cylinders with numbers 1–7.

methods. There is no single pattern or profile representing the entire GPCR family. And even if the query is predicted to belong to the GPCR super family, extreme divergence among families sometimes prevents further classification. In such cases, the query sequences are called "orphan" GPCRs. For example, the current GPCRDB ("Information system for G protein-coupled receptors (GPCRs)": http://www.gpcr.org/7tm/) contains about 4,650 GPCR entries, and more than 300 entries are designated as "orphans" or "putative/unclassified" GPCRs.

At varied degrees, such situations are shared with many other transmembrane proteins. We can expect that when we develop methods that can successfully classify this particular protein family, such methods can be applied easily for many other protein families. We therefore use the GPCR protein family in our study to evaluate performance of various classification methods.

## 2. DISCRIMINANT ANALYSIS

In Kim *et al.* (2000), we described a new method of protein classification that relies on neither multiple alignments nor pattern/profile database search. The new method uses a set of variables extracted from each protein sequence, and classifies them by using a "nonparametric" linear discriminant analysis. It is a linear discriminant analysis optimized with nonparametric "runs" criterion, instead of relying on parametric equations commonly used. This was because we wanted to avoid assuming any unreasonable statistical distribution. In this paper, we include other parametric and nonparametric discrimination methods and compare their performance for the GPCR family classification.

### 2.1. Input Variables

Instead of using multiple alignments, a set of variables extracted from each protein sequence is used in discriminant analyses. We used the same set of variables described in Kim *et al.* (2000). They include "amino acid index" and three periodicity statistics based on hydrophobicity and polarity.

In order to obtain "amino acid indices" from protein sequences, first a linear discriminant analysis is done based on 19 amino acid frequencies. The linear discriminant score (we call it "amino acid index") obtained from each protein sequence can be used as a single variable representing 19 amino acid frequencies. Note that using "amino acid index" instead of each amino acid frequency separately reduces the dimensionality from 19 to 1.

Other three variables are:

   i) Log of the average periodicity of the GES scale,
   ii) Log of the average periodicity of the polarity scale, and
   iii) Variance of first derivative of the polarity scale.

Distributions of the GES hydropathy index (Engelman *et al.*, 1986) and polarity along each protein sequence are examined using sliding window analysis (window size = 16 amino acids). And the "average periodicity" is calculated by counting how many times the property crosses over a neutral value ($-0.5$ for the GES and $8.325$ for the polarity) and normalized by protein lengths.

Figure 5 in the next page shows how these four input variables discriminate GPCRs from non-GPCRs. Note that although the figure shows only two-dimensional variable spaces, good discriminations between GPCR proteins (shown with "G" in the figure) and non-GPCR proteins (shown with "R" in the figure) were observed from any combination of the four variables.

## 2.2. Datasets

A training dataset containing 750 of known GPCR sequences (randomly sampled from GPCRDB) and 1,000 of non-GPCR sequences (randomly sampled from SWISS-PROT) were prepared. A smaller dataset including 100 each of GPCR and non-GPCR sequences were also prepared independently as a test dataset.

## 2.3. Discrimination Methods

In addition to the "nonparametric" linear discriminant analysis (nonparametric LDA) method described previously (Kim *et al.*, 2000), we included four other parametric and nonparametric discrimination methods:

   i) Linear discriminant analysis (LDA),
   ii) Quadratic discriminant analysis (QDA),
   iii) Logistic discriminant analysis (LOG), and
   iv) K-nearest neighbor method (KNN).

S-Plus statistical package with the MASS library (Venables and Ripley, 2002) was used except for the nonparametric LDA.

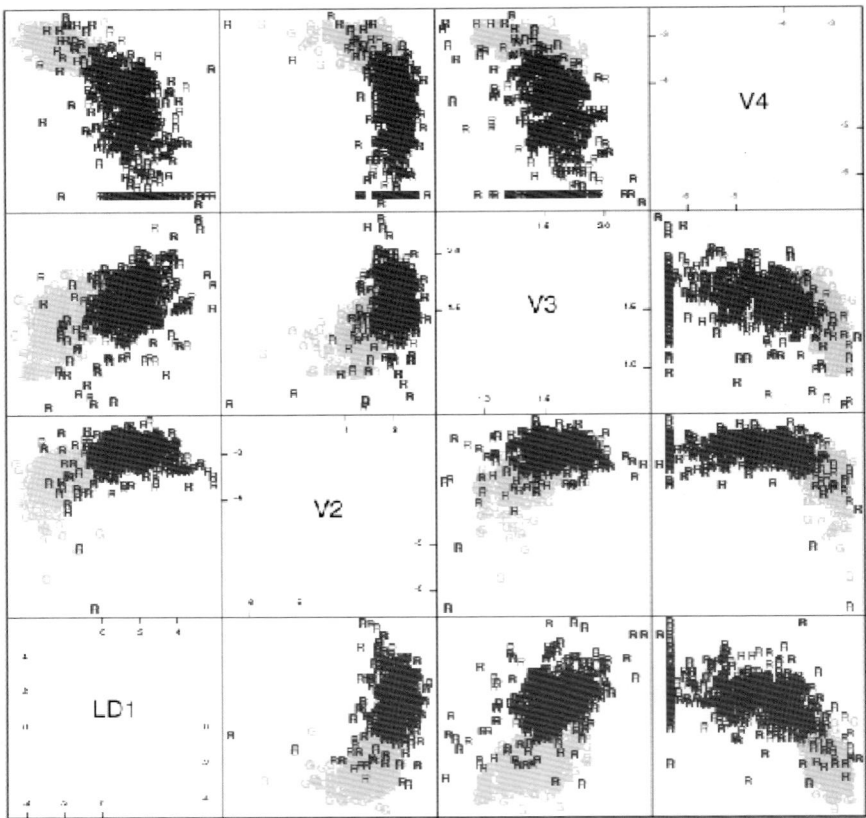

**FIGURE 5.** A multi-plot for the four input variables used in discriminant analyses. LD1: "amino acid index", V2, V3, and V4: hydrophobicity and polarity periodicity statistics. 1,750 proteins in the training dataset (described in the text) are plotted. "G" in grey color: GPCRs, "R" in black color: non-GPCR random proteins.

## 2.4. Performance Comparisons for the GPCR Protein Family Classification

Table 2 lists the results of GPCR classification by various discriminant analyses compared with other protein classification methods. Each method was trained on the training dataset described above, and their classification performance was tested on both of the training and test datasets. The cross-validation ("leave-one-out" test) was performed only for the three parametric discriminant analysis methods and nonparametric KNN method.

All of the four parametric and nonparametric methods performed similar to or better than previously described methods (PROSITE, Pfam, PRINTS, and nonparametric LDA) with % true positives higher than 98%. Surprisingly both nonparametric methods, KNN and nonparametric LDA, did not perform particularly

**Table 2**
**Performance Comparisons for the GPCR Classification.**

| Method | Against training dataset | | Against test dataset | | Cross-validation | |
|---|---|---|---|---|---|---|
| | % True + | % False + | % True + | % False + | % True + | % False + |
| LDA | 98.7 | 3.3 | 100 | 1 | 98.7 | 3.6 |
| QDA | 98.5 | 3.0 | 100 | 0 | 98.5 | 3.0 |
| LOG | 98.0 | 2.8 | 100 | 0 | 97.7 | 2.9 |
| KNN (k = 10) | 98.7 | 3.2 | 99 | 0 | 98.3 | 3.4 |
| Nonparametric LDA | 98.1 | 3.6 | 99 | 0 | — | — |
| PROSITE/pattern | 93.5 | 0.1 | 84 | 0 | — | — |
| PROSITE/profile | 98.8 | 0 | 94 | 0 | — | — |
| Pfam | 98.4 | 0.3 | 94 | 0 | — | — |
| PRINTS | 99.2 | 0.3 | 98 | 0 | — | — |

*Note*: "%True +": percent true positives, "% False +": percent false positives. For KNN analysis, k was varied between 5 and 20, but the performance with k = 5, 10, and 15 was similar. See Kim *et al.* (2002) for the PROSITE, Pfam, and PRINTS entries used in the analyses.

better than parametric discriminant analysis (DA) methods, although we cannot guarantee any assumption underlying parametric methods (*e.g.*, normal distribution, consistent covariance matrices). The false positive rates among DAs and KNN were also very similar to each other, but about 10 times higher than other methods (PROSITE, Pfam, and PRINTS). Classification of the independently prepared test dataset and the results of cross-validation were consistent with the classification results on the training dataset itself. As described before, classification by PROSITE/pattern search showed the lowest performance. It clearly indicates the limitation of searching short and less flexible regular expression patterns.

On the other hand, all of the DA and KNN methods outperformed other methods when tested on short sequences. Short subsequences (from 50 to 400 amino acids) were randomly sampled from the test dataset, and classification performance was compared among the nine methods listed in Table 2. Figure 6 in the next page compares % identifications by DA, KNN, and other methods. With test sequences with 50 or 75 amino acids in length, we could still obtain higher than 70% of positive identification by DA and KNN methods. In particular, nonparametric methods (KNN and nonparametric LDA) showed true positive rates higher than 80% if sequences were longer than 75 amino acids. On the contrary, Pfam, for example, could identify fewer than 50% of GPCR sequences when their length was 50 amino acids. Both PROSITE/pattern and PRINTS showed the lowest performance especially when sequences are short (300 amino acids or shorter). These observations were again consistent with the short patterns or fingerprints these classification methods use.

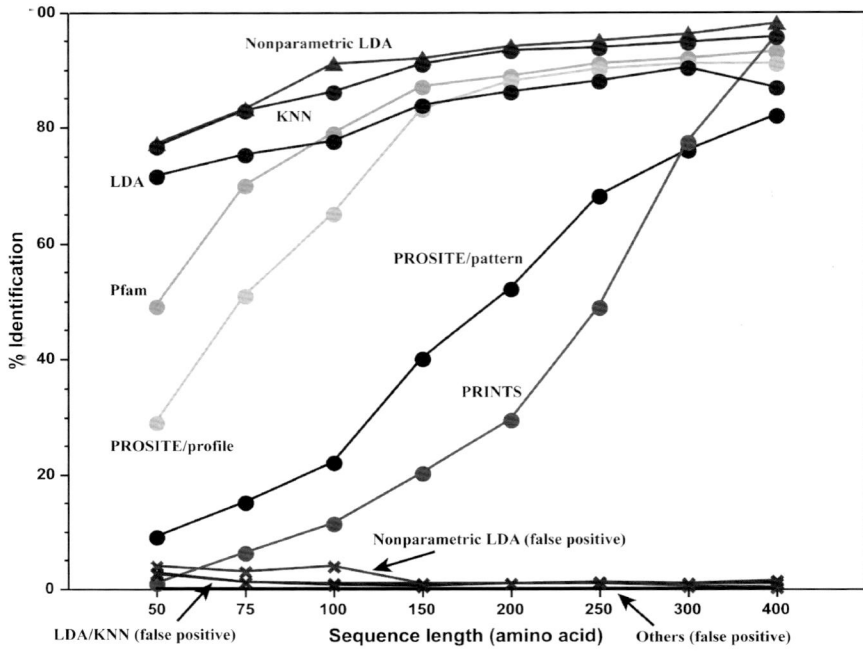

**FIGURE 6.** Performance comparison among classification methods against short protein sequences.

## 2. DEVELOPMENT OF A HIERARCHICAL CLASSIFICATION ALGORITHM

Our study showed that each protein classification method has different strength and weakness. All discrimination methods we examined (parametric or nonparametric) showed better performance even when protein sequences are extremely short. On the other hand, in general, these discrimination methods had much higher false positive rates. We should note that we simply relied on SWISS-PROT annotations to identify GPCRs in this study. Therefore, it is possible that there are some miss-identifications. Currently used methods (represented by PROSTIE, Pfam, and PRINTS in this study) have very low false-positive rates. However, they perform best when properties of query and training datasets are consistent, as shown in their weak performance against partial sequences. Therefore, these methods are not likely to identify new protein sequences if they are not closely related to any existing protein family or any existing protein family member. These methods also rely on the quality of multiple alignments among training sequences.

The ideal protein classification method should have reasonably low false positive rates but needs to be sufficiently flexible, so new types of proteins can be still identified or classified. In order to realize such optimal means of protein classification, we are currently developing an integrated algorithm that compliments weakness of various methods by combining various methods systematically and effectively.

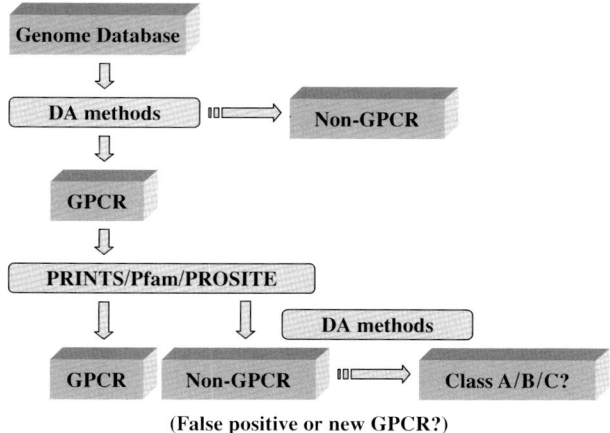

**FIGURE 7.** A hierarchical protein classification algorithm.

The flowchart shown in the next page presents a simple example of such hierarchical algorithm to identify potentially new GPCR sequences. In this hierarchical algorithm, more "non-specific" DA methods are first used to identify any possible candidate for GPCR proteins and discriminating them from "least likely to be GPCR" proteins. In the next step, other more stringent methods (*e.g.*, PROSITE, Pfam, and PRINTS) can be used to filter out "more likely to be GPCR" data. The remaining dataset could contain both actual "false positives" and also some new members of the GPCR family. Another level of DA or other clustering methods could be used to select such possible candidates and to perform more detailed classification.

Further examinations of various DA and other multivariate methods are required to identify which "non-specific" methods should be incorporated in this algorithm. It is also possible to create another hierarchical level among these "non-specific" methods to lower the false positive rates.

Our goal is to develop an integrated hierarchical algorithm that can take advantage from various protein classification methods. This algorithm can be applied for any kind of protein families, and incorporating DA and other flexible methods, we can apply this algorithm even for partial or short sequences as found in EST databases. It will be also useful to identify particular protein coding sequences from short fragments (*e.g.*, exons) from genomic data.


## 4. REFERENCES

Altschul, S.F., Gish, W., Miller, W., Myers, E.W., and Lipman, D.J., 1990, Basic local alignment search tool, *J Mol Biol.* **215**:403; http://www.ncbi.nlm.nih.gov/BLAST/.

Altschul, S.F., Madden, T.L., Schaffer, A.A., Zhang, J., Zhang, Z., Miller, W., and Lipman, D.J., 1997, Gapped BLAST and PSI-BLAST: a new generation of protein database search programs, *Nucleic Acids Res.* **25**:3389; http://www.ncbi.nlm.nih.gov/BLAST/.

Attwood, T.K., Blythe, M., Flower, D.R., Gaulton, A., Mabey, J.E., Maudling, N., McGregor, L., Mitchell, A., Moulton, G., Paine, K., and Scordis, P., 2002, PRINTS and PRINTS-S shed light on protein ancestry, *Nucleic Acids Res.* **30**:239; http://www.bioinf.man.ac.uk/dbbrowser/PRINTS/.

Bateman, A., Birney, E., Cerruti, L., Durbin, R., Etwiller, L., Eddy, S.R., Griffiths-Jones, S., Howe, K.L., Marshall, M., and Sonnhammer, E.L.L., 2002, The Pfam protein families database, *Nucleic Acids Res.* **30**:276; http://pfam.wustl.edu/index.html.

Falquet, L., Pagni, M., Bucher, P., Hulo, N., Sigrist, C.J., Hofmann, K., and Bairoch, A., 2002, The PROSITE database, its status in 2002. *Nucleic Acids Res.* **30**:235; http://www.expasy.ch/prosite.

Kim, J., Moriyama, E.N., Warr, C.G., Clyne, P.J., and Carlson, J.R., 2000, Identification of novel multi-transmembrane proteins from genomic databases using quasi-periodic structural properties., *Bioinformatics.* **16**:767.

Letunic, I., Goodstadt, L., Dickens, N.J., Doerks, T., Schultz, J., Mott, R., Ciccarelli, F., Copley, R.R., Ponting, C.P., and Bork, P., 2002, Recent improvements to the SMART domain-based sequence annotation resource, *Nucleic Acids Res.* **30**:242; http://smart.embl-heidelberg.de/.

Engelman, D.M., Steitz, T.A., and Goldman, A., 1986, Identifying nonpolar transbilayer helices in amino acid sequences of membrane proteins, *Annu. Rev. Biophys. Biophys. Chem.* **15**:321.

Venables, W.N., and Ripley, B.D., 2002, *Modern Applied Statistics with S.* Fourth Edition, Springer, New York.

*Chapter 10*

# Exploiting Natural Variation
# to Understand Gene Function in Pine

David B. Neale and Garth R. Brown

## 1. INTRODUCTION

Plant biology has entered into the functional genomics era. A finished genome sequence has been completed for the model dicot *Arabidopsis thaliana* (The Arabidopsis Genome Initiative, 2000) and is nearing completion for the model monocot *Oryza sativa* L. (Yu et al., 2002; Goff et al., 2002). The task of understanding the function of all 30,000± genes in these model plants has already begun. These model species possess all the desirable attributes for functional genomic analysis, e.g., rapid generation time, small genome size, well-characterized mutants and facile transformation. In *Arabidopsis*, genetic analyses are almost always performed on two major ecotypes, Columbia and Landsberg. Likewise in rice, a small number of varieties from two subspecies (*indica* and *japonica*) are most routinely used. The limitation that arises from concentrating functional genomic analyses to just a few genotypes is that the functional significance of naturally occurring allelic variation on plant phenotypes eludes understanding. From the perspective of the plant breeder, understanding of the function of a genetic locus is much less important than understanding the functional differences *among alleles at a locus*. To address questions of the functional significance of allelic variation, studies must be performed on a large number of diverse genotypes.

**David B. Neale**    Institute of Forest Genetics, University of California at Davis, Environmental Horticulture Dept., Davis, California 95616    **Garth R. Brown**    Institute of Forest Genetics, Pacific Southwest Research Station, USDA Forest Service, Davis, California 95616

*Genome Exploitation: Data Mining the Genome*, edited by J. Perry Gustafson, Randy Shoemaker, and John W. Snape.
Springer Science + Business Media, New York, 2005.

This leads to the application of association studies such as those frequently used in human genetics to identify disease-causing alleles for complex diseases (Risch, 2000; Cardon and Bell, 2001).

## 1.1. Association Mapping in Plants

The association mapping approach has just recently been applied to plants. Since the late 1980s, plant geneticists have used the quantitative trait locus (QTL) mapping approach to dissect complex traits in plants and in many cases have found markers linked to alleles of economic importance that can be used in marker-assisted breeding (Kearsey and Farquar, 1998; Paterson, 1998). However, the resolution of QTLs is poor (10–20 cM), making it a challenge to identify the gene(s) underlying QTLs. Positional cloning of QTLs has been accomplished in a few cases (Frary et al., 2000), but the task is enormous. A solution to this problem is to increase the amount of genetic recombination in the mapping population so that QTLs can be resolved to smaller chromosomal segments. The association mapping approach accomplishes this by taking advantage of historical recombination in a natural population as opposed to the small amount of recombination that occurs following a generation or two of crossing in standard QTL mapping populations (Lander and Schork, 1994).

The association mapping approach to dissecting complex traits and discovery of alleles for plant breeding will be applied to many crop plants in the near future (Rafalski, 2002). To date however, this approach has been used only in maize (Remington et al., 2001; Thornsberry et al., 2001). Due to the large size of the maize genome ($2.5 \times 10^9$ bp), it is currently not feasible to construct a high-density SNP (single nucleotide polymorphism) map such that a genome-wide scan can be performed to search for QTLs. The alternative is to use a candidate gene approach where SNPs in a select set of genes are associated with phenotypes. This approach assumes some a priori knowledge of the function of candidate genes. The first demonstration of association mapping in plants was by Thornsberry et al. (2001) who found that SNPs in the *Dwarf8* locus of maize associated with flowering time in a population of 92 maize inbred lines. One limitation of the association mapping approach is that false associations between SNPs and phenotypes are possible due to population structure and admixture. This possibility is likely in maize due to its domestication history. These questions were addressed in a companion study (Remington et al., 2001), where SSR markers were used to test for population structure. Population structure and admixture is likely to complicate association mapping in many of the inbred crops that have gone through severe domestication bottlenecks.

## 2. PINE IS A MODEL SPECIES TO UNDERSTAND ALLELIC EFFECTS ON PHENOTYPES

Pines (*Pinus*) are not generally thought of as good experimental plants for functional genomic analyses. Pines have long generation times, large genomes

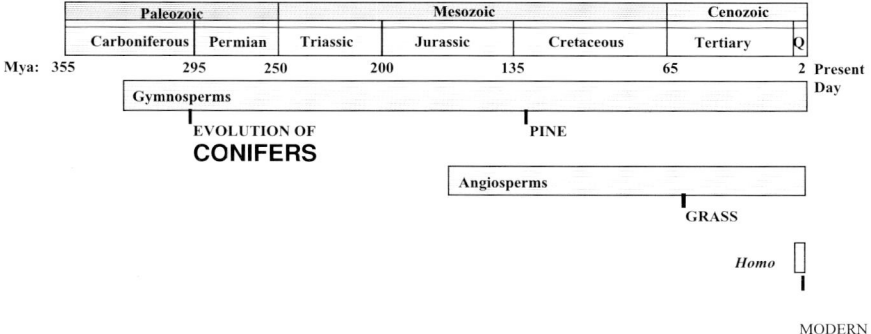

FIGURE 1. Gymnosperms, including pines, appeared much earlier in evolutionary history than angiosperms.

($1.0 \times 10^{10}$ bp), few characterized mutants and are often difficult to transform. In spite of these limitations, pines have a number of attributes that make them an excellent model for understanding the effects of allelic diversity on plant phenotypes. An association mapping approach is highly amenable to pine due to its evolutionary history, life history characteristics and reproductive biology.

## 2.1. Pines Are Evolutionarily Much Older than Crop Species

The gymnosperms evolved in the Carboniferous Period whereas the angiosperms did not appear until the Jurassic Period, more than 100 million years later (Figure 1). Pine first appeared during the Cretaceous and it was also the period of their major radiation. Thus, pines, of which there are more than 100 species, have been on earth for about 100 million years. Climatic conditions were very tropical during the Palaeocene and early Eocene that favored development and expansion of angiosperms and pushed pines into refugia. This period of fragmentation led to secondary centers of diversity and radiation, such as in Mexico.

There was a significant drop in temperatures at the end of the Eocene that caused widespread extinction of angiosperms and re-colonization of mid-latitudes by pines. The Pleistocene Epoch was a time of expansion and contraction of glaciers and thus the distributions of all flora. Pines forced into glacial refugia would have suffered losses of genetic diversity due to genetic drift. So even though pines may have passed through bottlenecks during their evolutionary history, many of the extant species have been on earth continuously for 100 million years or more, longer than most crop species.

## 2.2. Pines Are Mostly Undomesticated and Are Found in Large Random-Mating Populations

The domestication and breeding of pines is still in its infancy (Zobel and Talbert, 1984). The first attempts at genetic selection and breeding of pines did not begin until the mid-twentieth century, centuries later than the domestication of most agricultural crops. Therefore, pines have generally not suffered losses of diversity due to domestication. Many pine species have very large and often continuous populations. Some examples are Scots pine (*Pinus sylvestris* L.) in Europe and Asia and jack pine (*Pinus banksiana* Lamb.) and lodgepole pine (*Pinus contorta* Dougl.) in North America. However, pine populations in some parts of the world, notably around the Mediterranean, have been drastically reduced or altered due to over-harvesting and/or fire. The implications for association mapping are that in pine populations one might expect to find abundant allelic diversity that has resulted primarily from the forces of natural selection over a very long period of evolutionary time.

Pines have a mixed mating system (selfing and outcrossing) but most offspring result from outcrossing (Muona, 1990). Pollen and seed dispersal distances are also quite large. Collectively, these life history traits all lead to very large effective population sizes. The implication for association mapping is that individuals can be sampled from populations lacking large amounts of population structure and that these individuals will not be inbred.

### 2.3. Clonal Replication Permits Precise Evaluation of Phenotypes

The greatest experimental challenge in association mapping in humans is evaluation of the phenotype, which is always difficult in human genetic studies. Precise evaluation of phenotype is difficult in most animal systems due to problems in minimizing environmental variation. In plants, cloning or development of inbred lines is often possible which allows the establishment of replicated genetic tests. In pines, members of an association population can easily be cloned by methods such as somatic embryogenesis (SE) and rooted cuttings (Figure 2). SE can produce thousands of individuals per clone but is very costly. Rooted cuttings generate smaller numbers of individuals per clone but are much less expensive to produce. Once clones are produced it enables the evaluation of multiple phenotypes measured over time and space. Destructive sampling is possible on some individuals of the clone without losing the clone for future phenotypic evaluations.

Clonal tests can be established at multiple sites to test for G × E interactions and can also be measured over many years to address developmental questions. In summary, clonal replication of members of an association mapping population in pine enables precise and thorough evaluation of phenotypes that is not always possible in animal systems or even all crop plants.

### 2.4. Direct Determination of Haplotype

Statistical tests for association between genotype and phenotype can be performed using individual SNPs or based on the complete haplotype. Haplotype-based tests are considered to be more powerful. The problem is in how to determine or estimate the haplotype from zygotic data. In inbred crops, the problem can be avoided if individual plants are completely homozygous (Rafalski, 2002).

**FIGURE 2.** Clonal replication of loblolly pine by rooted cutting propagation.

However, using inbred material may not be optimal for association mapping because genetic diversity could be lacking.

A unique aspect of pine seed biology permits direct determination of haplotypes from individual trees. The pine seed endosperm (called megagametophyte) is haploid and results from the same meiotic event as the egg (Figure 3). DNA is easily isolated from the megagametophyte and DNA sequencing is performed on the haploid DNA templates to discover SNPs. By sequencing multiple megagametophytes from individual trees it is possible to determine both haplotypes. Alternatively,

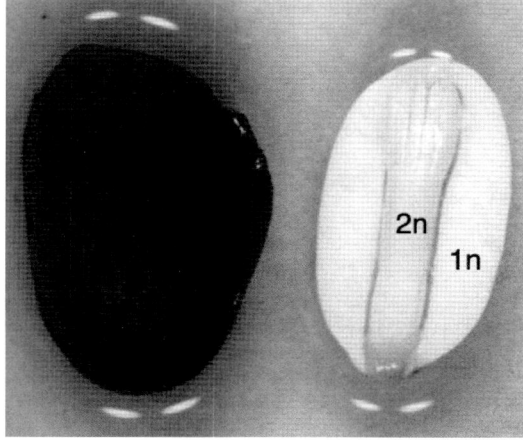

**FIGURE 3.** Pine seed showing 1n megagametophyte and 2n embryo.

diploid DNA can be sequenced in addition to one megagametophyte and the alternative haplotype can be inferred by subtraction. The implication for association mapping is that all haplotypes in a population can be determined directly, without error that is associated with haplotype estimation.

## 3. ASSOCIATION MAPPING IN LOBLOLLY PINE

Our ongoing research interest is in the genetic dissection of complex traits in forest trees. Nearly all traits of economic importance in forestry are quantitatively inherited, yet there is little understanding of the individual genes that control these traits. In loblolly pine (*Pinus taeda* L.), we seek to identify the genes controlling complex wood property traits. We have used the QTL mapping approach to identify chromosomal regions harboring QTL for both physical wood properties (wood density and microfibril angle) and chemical wood properties (percent lignin and cellulose) (Groover et al., 1994; Sewell et al., 2000, 2002; Brown et al., 2003). The mapping experiments have been repeated in different families and across different environments, therefore a fairly comprehensive picture has emerged as to the (1) number, (2) location and (3) size of effect of QTLs controlling these traits. However, the genes underlying these QTLs are still unknown. Positional cloning of QTLs in the large genome of loblolly pine is not feasible, therefore we have chosen to use a candidate gene-based association mapping approach toward identifying the genes underlying these QTLs.

### 3.1. Clonal Association Mapping Populations and Phenotypic Evaluations

A clonal association mapping population was assembled from 10 different clone banks and clonal seed orchards belonging to the Weyerhaeuser Company. The clone banks and seed orchards were located throughout the southeastern United States. Conscience effort was made to minimize genetic relatedness among individuals selected from the population. The trees varied in age but most were more than 15 years old. A sample of 425 clones were selected each with two copies (ramets) per clone (~850 trees in total).

Wood samples were taken from all trees by 5 mm increment cores. The wood samples were evaluated for several wood property traits including (1) wood specific gravity, (2) percent latewood, (3) microfibril angle and (4) percent lignin and cellulose. All of these traits had previously been mapped by QTL analyses.

### 3.2. SNP Discovery, Nucleotide Diversity and Linkage Disequilibrium

A list of candidate genes for the wood property traits was assembled (Table 1). These included (1) nine genes coding for enzymes in the phenylpropanoid pathway leading to the synthesis of lignin monomers (*pal, c4h-1, c4h-2, 4cl, c3h-2, ccoaomt, ccr, comt-2, cad*), (2) three genes coding for enzymes involved in supplying methyl

**Table 1**
**Estimates of Nucleotide Diversity for Chemical and Physical Wood Property
Candidate Genes in Loblolly Pine.**

| Gene | Locus | Length | $\theta_{TOTAL}$ |
|------|-------|--------|--------|
| phenylalanine ammonia-lyase | *pal-1* | 438 | 0.0034 |
| cinnamate 4-hydroxylase | *c4h-1* | 1718 | 0.0036 |
| 4-coumarate:CoA ligase | *c4h-2* | 912 | 0.0033 |
| | *4cl* | 1716 | 0.0051 |
| coumarate 3-hydroxylase | *c3h-2* | 1334 | 0.0010 |
| caffeoyl CoA O-methyltransferase | *ccoaomt* | 529 | 0.0060 |
| cinnamoyl CoA reductase | *ccr* | 616 | 0.0068 |
| caffeate O-methyltransferase | *comt-2* | 1275 | 0.0031 |
| cinnamyl alcohol dehydrogenase | *cad* | 391 | 0.0040 |
| S-adenosyl methionine synthetase | *sam-1* | 784 | 0.0038 |
| | *sam-2* | 463 | 0.0016 |
| glycine hydroxymethyltransferase | *glyhmt* | 573 | 0.0054 |
| LIM transcription factor | *ptlim1* | 428 | 0.0017 |
| | *ptlim2* | 453 | 0.0027 |
| cellulose synthase | *cesA3* | 998 | 0.0022 |
| arabinogalactan proteins | *agp-like* | 890 | 0.0020 |
| | *agp-4* | 402 | 0.0192 |
| | *agp-6* | 865 | 0.0103 |
| | Total | 14785 | 0.0042 |

groups to the phenylpropanoid pathway (*sams-1, sams-2, glyhmt*), (3) two genes
coding for transcription factors (*ptlim-1, ptlim-2*), (4) a gene coding for a cellulose
synthase (*cesA3*) and (5) three genes coding for arabinogalactan proteins (*agp-like,
agp-4, agp-6*). DNA sequence contigs could be found for all of the genes at the
loblolly pine xylem EST database (http://pinetree.ccgb.umn.edu/).

SNP discovery was performed by direct sequencing of PCR amplicons from
one or more fragments of each of the candidate genes. A panel of megagametophyte
DNA samples, one from each of 32 trees of the association mapping population,
was used for SNP discovery. Approximately 15 kb of 5′ untranslated, exon, intron
and 3′ untranslated sequence was searched for SNPs. In a preliminary analysis to
identify SNPs, sequences were aligned using the Sequencher software and SNPs
identified visually (Figure 4).

The frequency of SNPs in coding regions was 1/91 *versus* 1/37 in non-coding
regions (Table 2). The average frequency was 1/60 and is identical to that found
in one maize study (Ching et al., 2002), but less than that found in another study
(Tenaillon et al., 2001). SNP frequency in soybean (Zhu et al., 2003) is considerably
less than that in loblolly pine or maize (Table 2).

Estimates of average (coding and non-coding) nucleotide diversity in loblolly
pine varied more than 12-fold among genes (range = 0.0017 − 0.0192). There are
potentially many mechanisms responsible for differences among genes, nevertheless natural selection or genetic drift must be acting differentially among genes.

Haplotype

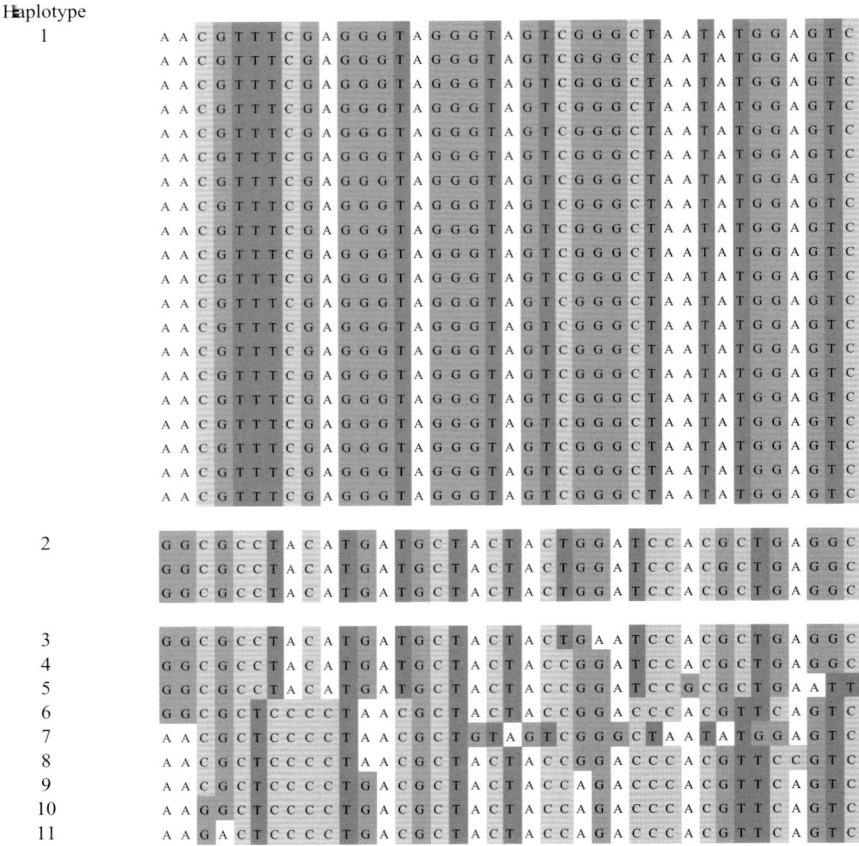

**FIGURE 4.** SNP haplotypes of the *4cl* gene among a sample of 32 loblolly pine megagametophytes.

The average nucleotide diversity in loblolly pine (0.0042) is somewhat less than that for maize but greater than that in soybean.

The extent of linkage disequilibrium (LD) is important to estimate as it determines the marker density that is needed to identify associations with phenotypes. LD varies among genes in loblolly pine but on average LD decays to $r^2 \leq 0.20$ within ~1000 bp (Figure 5). The limited estimated LD in loblolly pine is not unexpected due to its highly outcrossed mating system and evolutionary history. The implication for association mapping in loblolly pine is that genome-wide association studies would not be feasible at this time due to the high marker density that would be required and that although associations may be difficult to establish using a candidate gene approach, once established the marker should be physically very close to the functional polymorphism.

There are very few genome-wide studies of LD in plants, although one exception is that in *Arabidopsis thaliana*. Nordborg et al. (2002) found that LD decays

**Table 2**
**Estimates of Nucleotide Diversity in Plants**

| | Length | N | bp/SNP | | | $\Theta_{total}$ | Reference |
|---|---|---|---|---|---|---|---|
| | | | Coding | Non-coding | Total | | |
| loblolly | 14,785 | 32 | 91 | 37 | 60 | 0.0042 | This study |
| Scots | 2,045 | 20 | — | — | — | 0.0017 | Dvorynk et al., 2002 |
| maize | 6,935 | 36 | 124 | 31 | 60 | — | Ching et al., 2002 |
| maize | 14,420 | 25 | — | — | 27.6 | 0.0096 | Teniallon et al., 2001 |
| soybean | >76,000 | 25 | 503 | 283 | 308 | 0.00097 | Zhu et al., 2003 |

within ~250 kb in this selfing annual plant. Thus, genome-wide association mapping in *Arabidopsis thaliana* is probably quite feasible. In other plants, LD has been estimated within individual genes or at specific chromosomal regions. Remington et al. (2001) also found that LD was variable among genes in maize but on average decays within ~1500 bp. Tenaillon et al. (2001) found that LD decayed even more rapidly in a more diverse set of maize accessions (~100–200 bp). In soybean, LD is more like that in *Arabidopsis*, decaying only within ~50 kbp. Clearly, LD varies among species depending on their evolutionary histories as well as among genes or regions of the genome within species. The choice of population within which to estimate LD will also greatly affect estimates.

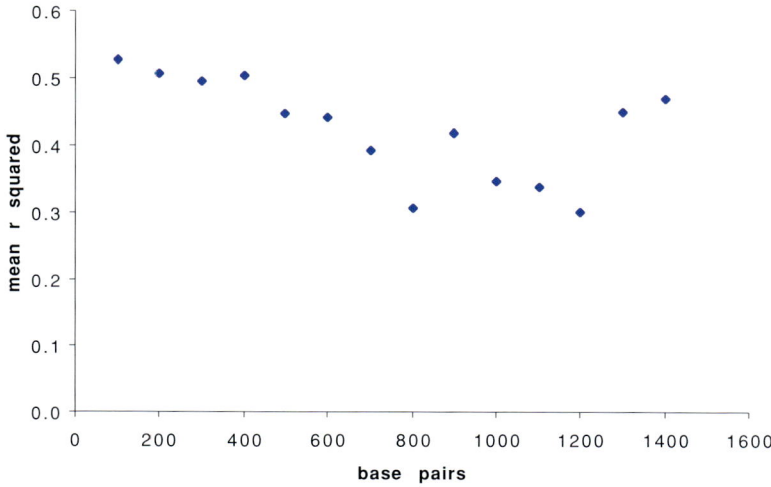

**FIGURE 5.** Average rate of decay of linkage disequilibrium ($r^2$) in loblolly pine.

**Table 3**
**Association Tests of SNP Genotype with Wood Property phenotype in loblolly pines**

| GENE | SNP | PHENOTYPE | F | Pr > F |
|------|-----|-----------|---|--------|
| 4cl | C/G | ewsg (11–15) | 4.11 | 0.043 |
| | | %lw (3–5) | 6.16 | 0.014 |
| | | %lw (11–15) | 9.72 | 0.002 |
| c4h-1 | C/A | ewsg (3–5) | 3.55 | 0.029 |
| | C/T | ewsg (3–5) | 3.04 | 0.049 |
| | | lwsg (11–15) | 3.02 | 0.050 |
| comt-2 | C/T | ewsg (11–15) | 4.20 | 0.041 |
| | C/T | ewsg (11–15) | 6.61 | 0.011 |
| | | emfa (11–15) | 3.90 | 0.049 |
| agp-like | C/A | %lw (11–15) | 4.19 | 0.016 |

## 3 3. Associations between SNP Genotype and Phenotype

We have identified a total of 245 SNPs among the 18 wood property candidate genes. A subset of the 50 most informative and potentially functional SNPs were typed in the full set of 425 clones from the association mapping population. SNP genotyping was performed using the Template-directed Dye-terminator Incorporation and Florescence Polarization detection (FP-TDI) assay (Gill et al., 2003). Wood property phenotypes were also evaluated for all 425 clones. Estimates are based on the mean of two ramets per clone. A statistical adjustment was performed to account for differences among clonal means due to test site differences.

ANOVA (SAS) was used to test for statistical association between 15 SNPs and 8 wood property phenotypes (120 tests in total). Those tests with $P \leq 0.05$ are shown in Table 3. These are preliminary results and no adjustments were made to account for multiple testing. The C/G SNP found in the 4cl gene was associated with average earlywood specific gravity in rings 11–15 and with average percent latewood in both rings 3–5 and rings 11–15. QTLs for both earlywood specific gravity and percent latewood map near the 4cl gene on linkage group 7 (Brown et al., 2003). Both the C/A SNP and the C/T SNP found in the c4h-1 gene were associated with earlywood specific gravity. A QTL for earlywood specific gravity was found at the same map position as the c4h gene on linkage group 3 (Brown et al., 2003). The two C/T SNPs found in the comt-2 gene also associated with earlywood specific gravity and a QTL is located fairly close to comt-2 on linkage group 11. Although preliminary, these results suggests that allelic variation in genes coding for enzymes in the phenylpropanoid pathway may be responsible for differences in the density of wood in loblolly pine. These findings await further testing and verification, but do suggest that association mapping can be applied to pine and that alleles can be discovered that may have significant economic value.

## 4. SUMMARY

Plant and animal breeders have made tremendous progress using phenotypic selection and quantitative genetic theory. Genetically improved varieties have resulted without any knowledge of the specific genes affecting desirable phenotypes. Genomic technologies now make it possible to identify the loci affecting phenotypes and measure the relative effects of different alleles. Breeders will soon have the option to select directly on genotype. However, there will be considerable challenges in accurately estimating the effects of many different alleles on complex trait phenotypes. Because of their evolutionary history, life history and reproductive characteristics, pines provide an excellent model system to begin to establish relationships between genotype and phenotype. Knowledge gained from pines will not only be used in applied forest tree breeding but might also serve for basic discovery that can be transferred to other plant and animal systems.

## 5. REFERENCES

Brown, G.R., Bassoni, D.L., Gill, G.P., Fontana, J.R., Wheeler, N.C., Megraw, R.A., Davis, M.F., Sewell, M.M., Tuskan, G.A., and Neale, D.B., 2003, Identification of quantitative trait loci influencing wood property traits in lolloby pine (*Pinus taeda L.*) III. QTL verification and candidate gene mapping, *Genetics* (in press).

Cardon, L.R., and Bell, J.I., 2001, Association study designs for complex diseases, *Nat. Rev. Genet.* **2**: 91–99.

Ching, A. et al., 2002, SNP frequency, haplotype structure and linkage disequilibrium in elite maize inbred lines, *BMC Genetics* **3**: 19–32.

Frary, A., Nesbitt, T.C., Grandillo, S., van der Knaap, E., Cong, E., Liu, J.P., Meller, J., Elber, R., Alpert, K.B., and Tanksley, S.D., 2000, fw2.2: A quantitative trait locus key to the evolution of tomato fruit size. *Science* **289**: 85–88.

Gill, G.P., Brown, G.R., and Neale, D.B., 2003, A sequence mutation in cinnamyl alcohol dehydrogenase gene associated with altered lignification in loblolly pine, *Plant Biotech. J.* (in press).

Goff, S.A. et al., 2002, A draft sequence of the rice genome (*Oryza sativa L.* ssp.*japonica*), *Science* **296**: 92–100.

Groover, A.T., Devey, M.E., Fiddler, T.A., Lee, J.M., Megraw, R.A., Mitchell-Olds, T., Sherman, B.K., Vujcic, S.L., Williams, C.G., and Neale, D.B, 1994, Identification of quantitative trait loci influencing wood specific gravity in loblolly pine, *Genetics* **138**:1293–1300.

Kearsey, M.J., and Farquhar, A.G.L., 1998, QTL analysis in plants; where are we now?, *Heredity* **80**: 137–142.

Lander, E.S., and Schork, N.J., 1994, Genetic dissection of complex traits, *Science* **265**: 2037–2048.

Muona, O., 1990, Population genetics in forest tree improvement, in: *Plant Population Genetics, Breeding, and Genetic Resources*, A.H.D. Brown, M.T. Clegg, A.L., Kahler, B.S. Weir, eds., Sinauer Associates Inc., Sunderland, pp. 282–298.

Norborg, M., Borevitz, J.O., Bergelson, J., Berry, C.C., Chory, J., Hagenblad, J., Kreitman, M., Maloof, J.N., Noyes, T., Oefner, P.J., Stah, E.A., and Weigel, D. 2002, The extent of linkage disequilibrium in *Arabidopsis thaliana*, *Nat. Genet.* **30**: 190–193.

Paterson, A.H., 1998, *Molecular Dissection of Complex Traits*, CRC Press, New York.

Rafalski, A., 2002, Applications of single nucleotide polymorphisms in crop genetics, *Curr. Opin. Plant Biol.* **5**: 94–100.

Remington, D.L., Thornsberry, J.M., Matsuoka, Y., Wilson, L.M., Whitt, S.R.., Doebley, J., Kresovich, S., Goodman, M.M., and Buckler IV, E.S., 2001, Structure of linkage disequilibrium and phenotypic associations in the maize genome. *Proc. Natl. Acad. Sci.* **98**: 11479–11484.

Risch, N.J., 2000, Searching for genetic determinants in the new millennium, *Nature* **405**: 847–856.

Sewell, M.M., Bassoni, D.L., Megraw, R.A., and Wheeler, N.C., 2000, Identification of QTLs influencing wood property traits in loblolly pine (*Pinus taeda* L.). I. Physical wood properties, *Theor. Appl. Genet.* **101**:1273–1281.

Sewell, M.M., Davis, M.F., Tuskin, G.A., Wheeler, N.C., Elam, C.C., Bassoni, D. L., and Neale, D.B., 2002, Identification of QTLs influencing wood property traits in loblolly pine (*Pinus taeda* L.) II. Chemical wood properties, *Theor. Appl. Genet.* **104**(2–3):214–222

Tenaillon, M.I., Sawkins, M.C., Long, A.D., Gaut, R.L., Doebley, J.F., and Gaut, B.S., 2001, Patterns of DNA sequence polymorphism along chromosome 1 of maize (*Zea mays* ssp. *mays L.*), *Proc. Natl. Acad. Sci.* **98**: 9161–9166.

The Arabidopsis Genome Initiative, 2000, Analysis of the genome sequence of the flowering plant *Arabidopsis thaliana*, *Nature* **408**:796–815.

Thornsberry, J.M., Goodman, M.M., Doebley, J., Kresovich, S., Nielsen, D., and Buckler IV, E.S., 2001, *Dwarf8* polymorphisms associate with variation in flowering time, *Nat. Genet.* **28**: 286–289.

Yu, J., et al., 2002, A draft sequence of the rice genome (*Oryza sativa L.* ssp. *Indica*), *Science* **296**: 79–92.

Zhu, Y.L., Song, Q.J., Hyten, D.L., Van Tassell, C.P., Matukumalli, L.K., Grimm, D. R., Hyatt, S.M., Fickus, E.W., Young, N.D., and Cregan, P.B., 2003, Single-nucleotide polymorphisms in soybean, *Genetics* **163**: 1123–1134.

Zobel, B.J., and Talbert, J.T., 1984, *Applied Forest Tree Improvement*, John Wiley & Sons, New York.

## Chapter 11

# Merging Analyses of Predisposition and Physiology Towards Polygene Discovery

Daniel Pomp, Mark F. Allan, and
Stephanie R. Wesolowsk

## 1. INTRODUCTION

Most quantitative traits are exceptionally complex, with relatively equal contributions of genetic susceptibility and interacting environmental factors. Predisposition to a phenotypic range for a complex trait such as body weight results from combinations of relatively small effects of DNA variations within a large number of unidentified polygenes, known as quantitative trait loci (QTL). Over 200 QTL have been reported for growth and body composition traits in the mouse, likely representing at least 50 to 100 distinct genes (Figure 1). While molecular biology has yielded significant gains in understanding complex traits such as weight regulation at the metabolic and physiological levels (e.g. leptin, melanocortin and insulin pathways), the genetic architecture of obesity predisposition remains essentially undefined. This large gap between our extensive knowledge of physiological mechanisms underlying body weight, and our embryonic understanding of how genetic predisposition is manifested, impairs identification of genes underlying relevant QTL and inhibits gene-based development of diagnostic and therapeutic tools.

We propose a central hypothesis that the majority of genes controlling predisposition to complex traits such as body weight and obesity are involved in

**Daniel Pomp, Mark F. Allan, and Stephanie R. Wesolowsk**    A218 Animal Science, University of Nebraska, Lincoln, Nebraska 68583-0908

*Genome Exploitation: Data Mining the Genome*, edited by J. Perry Gustafson, Randy Shoemaker, and John W. Snape.
Springer Science + Business Media, New York, 2005.

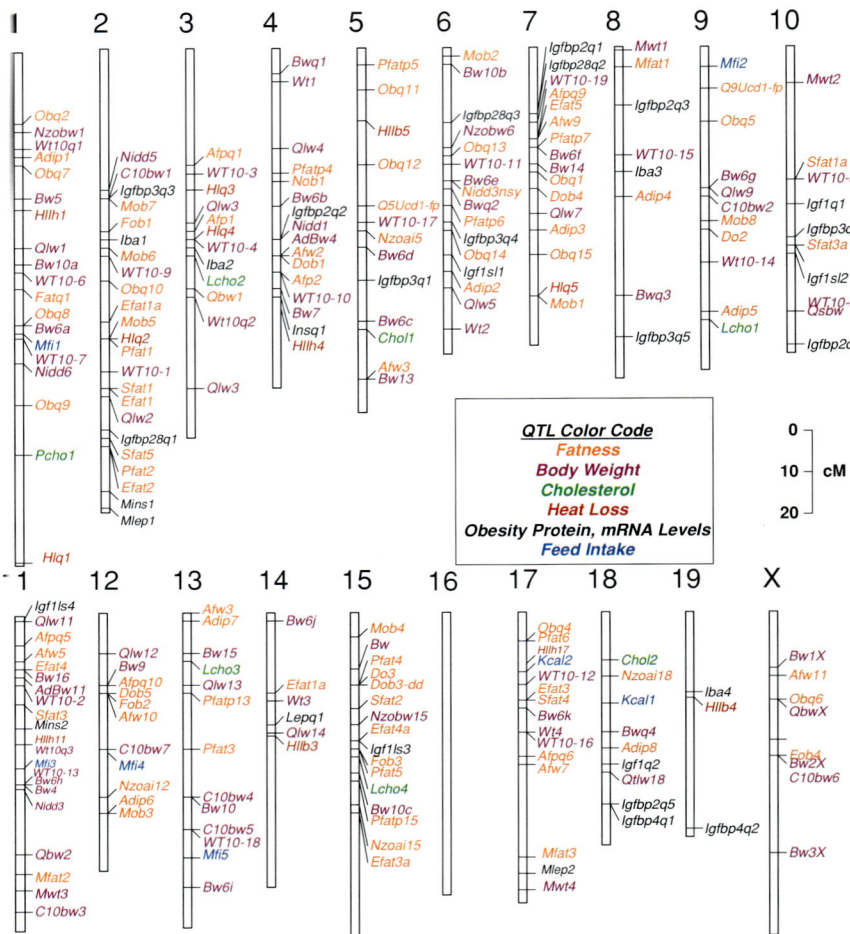

**FIGURE 1.** Mouse Body Weight Predisposition Map. The current status of mapped quantitative trait loci (QTL) for fatness (orange), body weight and weight gain (violet), serum cholesterol (green), heat loss (red) and levels of mRNA or proteins for candidate genes with physiological relevance to obesity (black). All body weight and weight gains are from animals of 6 weeks of age or older. Lengths of chromosomes and QTL map positions are according to the Mouse Genome Database. It is emphasized that QTL map positions may be inherently inaccurate (in some cases 95% confidence intervals for a QTL include a majority of a chromosome. Only QTL mapped with at least genome-wide 5% significance levels were included in the map, but several suggestive QTL were included when their map positions reaffirmed map positions of other QTL from independent experiments. Symbols for QTL are as presented in the literature (see Elo (2003) for full list of references). If authors did not provide symbols, those suggested by Chagnon et al. (2003) were used. Adapted from Elo (2003).

trans-regulation of the primary physiological pathways directly regulating energy balance phenotypes. This hypothesis has been formulated based on several areas of accumulated data. First, few causative mutations have been found within energy balance candidate genes despite significant detection efforts in humans

(Chagnon et al., 2003). Second, studies localizing QTL regulating mRNA or protein levels of such candidate genes have primarily identified trans-acting regulation. And third, recent genome-wide evaluations have found that trans-acting loci are the primary drivers of variation in gene expression in yeast (Brem et al., 2002; Yvert et al., 2003), *Drosophila* (Montooth et al., 2003) and mice (Schadt et al., 2003a). We contend that the paradigm of "Quantitative Genomics", whereby large-scale sub-phenotyping at the transcriptional, proteomic and metabolomic levels is performed within the context of a QTL mapping population, will be a powerful force in dissecting the genetic architecture of complex trait predisposition.

## 2. SIGNIFICANCE

Carcass and body composition traits constitute extremely important considerations of modern livestock production systems where consumer health concerns and marketing perspectives play increasingly prominent roles. The problem of excess fat in livestock and poultry carcasses is ubiquitous and has serious consequences for the animal industry at four levels: health perceptions of consumers; wasteful production of an undesired biological component; labor costs associated with trimming waste fat; and lower biological efficiencies of fatter animals (Eisen, 1989).

An estimated 65% of U.S. adults are overweight, and 31% are obese, with higher percentages in female minority populations (Flegal et al., 2002). Overweight and obese conditions substantially increases risk of hypertension, dyslipidemia, type 2 diabetes, coronary heart disease, stroke, gallbladder disease, sleep apnea, osteoarthritis and respiratory problems, and endometrial, breast, prostate, and colon cancers, combining to form the single largest cause of death in developed countries (NHLBI, 1998).

Understanding the roles of specific loci in genetic susceptibility to obesity is critical to improving human health and quality of life. As Comuzzie and Allison (1998) stated, "One of the greatest challenges in biomedical research today is the elucidation of the underlying genetic architecture of complex phenotypes such as obesity." Unfortunately, little progress has been made despite significant research attention. A tremendous gap exists between our embryonic knowledge of the nature of genetic predisposition to obesity and our bourgeoning understanding of its physiological and molecular underpinnings. Given the ~75 billion dollar annual health costs associated with obesity (Finkelstein et al., 2003) finding novel targets for pharmacological intervention and pharmacogenomic management is critical. Enhanced understanding of how complex traits are controlled will also aid in elucidating the nature of predisposition for other diseases, including certain forms of cancer, and will be broadly applicable to many important phenotypes.

## 3. BACKGROUND

A significant heritable component for disease was formally recognized as long ago as the turn of the 20th century, when Sir Archibald Garrod (1902) described the heritable nature of alkaptonuria. Since then, well over 5,000 genetic disorders exhibiting Mendelian inheritance have been described and the molecular basis for many of these has been identified. These diseases have phenotypes, which fall into discrete classes (e.g. afflicted vs. not afflicted). In contrast, most economically relevant traits in animal agriculture, and most common human maladies such as obesity, exhibit continuous phenotypic variation and a predominantly multifactorial and polygenic basis (Festing, 1979; Rich, 1990). While certain rare mutations have been identified accounting for a small minority of extreme phenotypes (e.g., Animals: McPherron and Lee, 1997; Humans: Chagnon et al., 2003), the actual identity of genes segregating and contributing to common phenotypes in populations is essentially unknown.

Location of polygenic factors controlling inheritance of quantitative traits can be established by tracking the segregation of closely linked markers (QTL mapping). This dates back to early work by Sturtevant (1913) with *Drosophila*, Haldane et al. (1915) with mice and Sax (1923) with *Phaseolus vulgaris*. For the subsequent 70 years, analyses continued to use visible phenotypic markers and protein variants. However, DNA marker analysis of quantitative traits has recently gained prominence with development of ubiquitous and polymorphic marker systems (e.g. Dietrich et al., 1996) and powerful statistical methodologies (e.g. Lander and Botstein, 1989; Zeng, 1993; Zeng, 1994; Haley et al., 1994). The first extensive use of DNA markers was in plants, where large suitable families could be generated (Paterson et al., 1988; Martin et al., 1989). The feasibility of performing QTL mapping in mammals was first demonstrated using rats (Hilbert et al., 1991; Jacob et al., 1991), but the mouse has subsequently dominated research in polygenic analysis including body weight regulation (Brockmann and Bevova, 2002).

Over 200 QTL for growth and obesity-related traits have been localized in the mouse, representing a comprehensive and exhaustive portrait of the predisposition map for mouse body weight regulation. For simplicity and convenience, we have graphically summarized the ~50 papers contributing this information in Figure 1 (see also Table 3 in each of Rocha et al., 2004a,b). However, few (if any) of these QTL have been unequivocally cloned, dramatically limiting the ability to harness this critical mass of information for the betterment of human health. This scenario is all the more surprising due to the fact that advancements in knowledge on the endocrine, biochemical and molecular underpinnings of obesity represent a significant success story in modern biology, sparked originally by study of spontaneous mutations causing obesity (e.g. Zhang et al., 1994; Michuad et al., 1994; reviewed by Chua, 1997) and later by targeted single gene mutations in a multitude of genes with relevance to energy balance (see reviews by Bray and Tartaglia, 2000; Barsh and Schwartz, 2002). This frustrating gap in our knowledge of physiological mechanisms underlying obesity, and the nature of genetic predisposition to obesity, was insightfully summarized by George Bray and Claude Bouchard (1997), who wrote: "Unfortunately, the spectacular gains in understanding the biology of energy balance of the last few years have not yet translated into significant advances on the genetic front. This is particularly striking when one realizes that so far there is not one single obese human being whose excess body fat can be explained by a specific mutation in one of the genes exerting its effects in relevant energy balance pathways".

In recent years there have been a few reports of single gene mutations resulting in human obesity (e.g., Farooqi et al., 2003; reviewed by Chagnon et al., 2003), and we have seen a somewhat promising trend in the ability to elucidate the underlying identities of QTL including those underlying obesity-related traits (see review by Korstanje and Paigen, 2002). However, the painstaking and difficult nature of the process bodes poorly for rapid progress (Nadeau and Frankel, 2000). This dilemma grows in parallel with the continued rapid pace of discovery in understanding energy balance mechanisms (e.g. Schwartz et al., 2003; Wang et al., 2003b).

## 4. TRANSCRIPTOME MAPPING: A NEW EXPERIMENTAL PARADIGM FOR ANALYSIS OF COMPLEX TRAIT GENETICS

A new paradigm for bridging the gap between our knowledge of the physiology and predisposition of obesity is the combining of QTL mapping with large-scale gene expression analysis. Transcriptome mapping (Williams et al., 2002b), also called "genetical genomics" (Jansen and Nap 2001; Jansen 2003), treats gene expression levels of any particular gene measured across different individuals as an expression-level polymorphism that in principle reflects the underlying genetic variation (Dumas et al., 2000; Jansen and Nap, 2001; Doerge, 2002). This type of analysis was actually pioneered by Damerval et al. (1994) and de Vienne et al. (1994) using proteomic evaluation (and later extended by the same group to the transcriptome (Consoli et al., 2002)) in an $F_2$ population of maize (*Zea mays* L.). Transcriptome mapping has been highlighted as a powerful mechanism to dissect complex traits and make more efficient the selection of candidate genes underlying predisposition loci, with recent successful implementation in yeast (Brem et al., 2002; Yvert et al., 2003), *Drosophila* (Wayne and McIntyre, 2002; Montooth et al., 2003) and mice (Hitzemann et al., 2003; Schadt et al., 2003a,b).

A critically important question that transcriptome mapping promises to help answer regards the underlying nature of obesity QTL. Are primary obesity QTL represented by sequence variation within genes with major roles in energy balance pathways or, as first proposed several years ago (Pomp, 1999), do they regulate such genes in a trans-acting manner? Although a plethora of candidate gene analyses in humans has provided mixed results (e.g. MC3R: Farooqi et al., 2003; UCP2: Esterbauer et al., 2001; LEPR: Heo et al., 2001), the latter speculation is strongly supported when the transcriptome/proteome mapping paradigm has been applied on a modest basis by examining one or a few transcripts/proteins at a time. For example, mapping for determinants of levels of plasma leptin, IGF-1 and IGF-1 binding proteins has identified multiple QTL for each trait, but none coinciding with map positions of the structural genes themselves in mice (e.g. Mehrabian et al., 1998; Brockmann et al., 2000; Rosen et al., 2000) or humans (Hixson et al., 1999). Koza et al. (2000) identified several trans-acting QTL for induced *Ucp1* mRNA levels in mice, and we have found similar results for control of hypothalamic mRNA levels of *Rpl3*, *Oxt* and *Timp2* (Wesolowski et al., 2003; Wesolowski and Pomp, Unpublished Data). In pigs, Rohrer et al. (2001) identified only trans-acting QTL influencing plasma FSH levels.

Although these very limited findings provide a tantalizing glimpse into the nature of complex trait predisposition and the underlying QTL, transcriptome mapping offers the potential for several orders-of-magnitude greater power and resolution. In yeast, for example, Brem et al. (2002) found QTL for 570 expressed genes in a cross between laboratory and wild strains. Of these, 36% were apparently due to polymorphisms within the gene themselves, while the remainder was controlled by a small group of trans-acting modulator loci each regulating from 7 to 94 genes of related function. Similar findings were reported by Schadt et al. (2003a) in their mouse study where they further found that the stronger the evidence for an expression QTL, the more likely it would map within the structural gene itself. This latter result is expected as DNA mutations within a gene and affecting the expression of that gene should be easier to identify than second-order effects. However, it should be noted that in such cases, the causal polymorphism may be closely linked to, but not part of, the gene whose expression is being evaluated. Also, such results could possibly be artificially be created by polymorphisms influencing hybridization efficiency, although this would be more likely when shorter oligonucleotide probes are employed in a microarray.

Recently, Yvert et al. (2003) determined that most gene expression differences in a cross between laboratory and wild strains of yeast mapped to trans-acting loci. Furthermore, trans-regulatory variation was broadly dispersed across different classes of genes with a wide variety of functions. And further compelling evidence that obesity QTL may be represented by sequence variation within trans-acting loci that regulate genes with major roles in energy balance pathways, is provided by the recent study of Montooth et al. (2003) using *Drosophila*. In that experiment, trans-regulatory variation was found in metabolic enzyme activity for each key determinant of metabolism and respiration measured.

The study by Schadt et al. (2003a) specifically targeted dietary-induced obesity in mice and identified two promising candidates potentially representing the MMU2 QTL described by Lembertas et al. (1977). This distal region of mouse chromosome 2 is one of the most relevant to obesity predisposition in the mouse genome (see Figure 1 and Chagnon et al., 2003). Not only is this region well populated with multiple body weight and fatness QTL, from crosses employing different approaches and genetic backgrounds (Lembertas et al., 1997; Mehrabian et al., 1998; Rocha et al., 2003a,b), QTL harbored in this region have among the largest effects of any body weight and obesity polygenes ever localized (Pomp, 1997; Rocha et al., 2003a,b).

In regard to regulation of body weight and obesity, critical questions remain to be answered after these pioneering, yet quite preliminary, initial studies using the transcriptome mapping paradigm. For example, will expression QTL represent genes with primary (cis) or secondary (trans) roles in energy balance? A second important question to address is whether expression QTL directly underlie obesity QTL? And more globally, can the transcriptome mapping paradigm indeed create a more successful environment for routine cloning of obesity predisposition genes?

## 4.1. Expected Outcomes From the Transcriptome Mapping Paradigm

Establishment of a robust system that can genetically map loci that modulate the steady-state levels of any gene showing transcript level variation between populations with divergent phenotypes is expected to provide a wealth of information regarding genetic architecture of complex traits. A hypothetical snap-shot of the types of outcomes this research paradigm can provide is summarized in Figure 2, where the X-axis represents the known genetic map position of each gene represented on a large-scale expression microarray, and the Y-axis represents the estimated genetic map location of the single QTL that explains the most variation in expression levels for each gene on the X-axis. Based on the early transcriptome mapping efforts, three patterns of transcript regulation are being revealed, with a fourth pattern possible.

First, a large number of genes plotted along the diagonal will suggest that their transcript levels are cis-regulated (i.e. the location of their QTL transcript regulators genetically map to the physical location of the genes themselves). We would speculate that these are due to promoter polymorphisms or other variants within the genes themselves that affect transcript level. This "cis-diagonal" (Williams et al., 2002b) can immediately uncover high-quality candidate genes potentially representing QTL for growth and obesity phenotypes, especially when the map position of the expression QTL falls under the QTL peak for the end-point phenotype (e.g., weight, fat, intake). A second class of genes would be those controlled by unlinked trans-regulators. These will be evidenced by the genetic locations of the controlling QTLs being different than the physical location of the genes they regulate. Evaluation of the trans-regulating patterns of expression phenotypes will add immense value to selection of candidate genes representing QTL, by implicating pathways and mechanisms underlying the mechanism of action of the QTL. The third, and potentially most interesting class of genes, is QTL transcript modulators that would regulate the steady-state transcript levels of tens or even hundreds of genes spread across the genome. These master transcript modulators would be identified by the horizontal strips of plotted QTL modulators, indicating the presence of one or a few tightly linked regulatory genes. This class of results will represent two important findings. First, the QTL may be in a key gene within a pathway that, when perturbed by a polymorphism, causes a cascade of effects that are evidenced by multiple expression changes in other genes. Second, the QTL may represent a key genetic control switch such as a transcription factor or helicase, a polymorphism within which could cause a multitude of changes in expression of genes throughout the genome. A speculative fourth class of genes would be those representing potential "expression neighborhoods" (Oliver et al., 2002; Spellman and Rubin, 2002), although evidence for such results has not yet been observed in the transcriptome mapping experiments conducted to date.

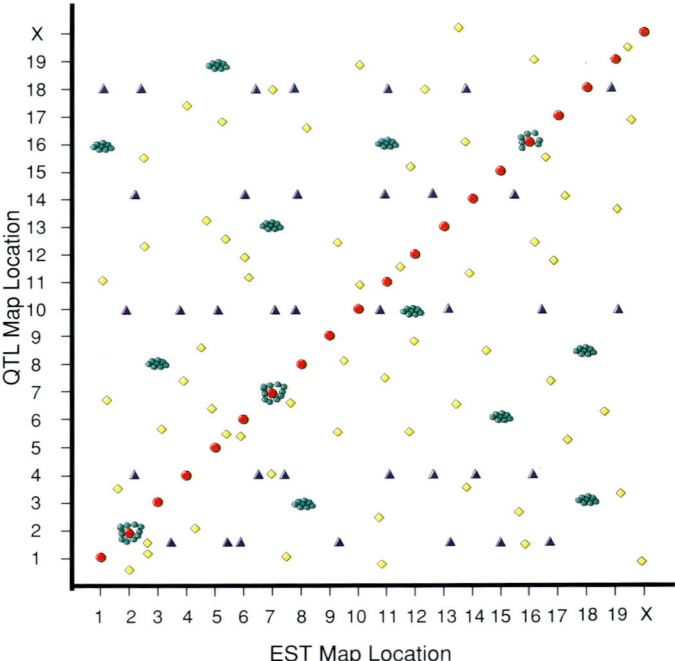

FIGURE 2. Hypothesized results of mRNA expression profiling across a genotyped quantitative trait locus (QTL) mapping population. The X-axis is the general map position of each expressed sequence tag (EST) or gene in the expression array. The Y-axis is the general map position of the QTL that explains the most variation in expression levels of each EST or gene in the expression array. Four generalized scenarios are described. Predominantly, levels of expressed genes will be controlled by trans-acting QTL (scattered yellow diamonds). Cis-acting QTL (diagonal red circles) may represent genetic variation within the regulatory or coding regions of the expressed genes themselves. In addition, a single QTL may result in changes in expression levels of many unlinked genes (horizontal blue triangles), either due to direct pleiotropy or to multiple changes in a regulatory cascade resulting from alternation of expression in a single key gene. Finally, clusters (small green dots) of gene expression changes could result from changes due to linkage of multiple expressed genes to a single regulatory QTL, or alternatively as a result of coordinated expression neighborhoods. Special thanks and acknowledgement to Rob Williams, David Threadgill and colleagues for useful discussions.

We expect these efforts to begin to enable development of an initial framework for understanding the genetic architecture of obesity predisposition. Such studies should greatly facilitate testing our hypothesized structure for such architecture (Figure 3), whereby genes controlling predisposition to complex traits such as body weight and obesity are, for the most part, involved in trans-regulation of the primary physiological pathways directly regulating phenotypes involved in energy balance.

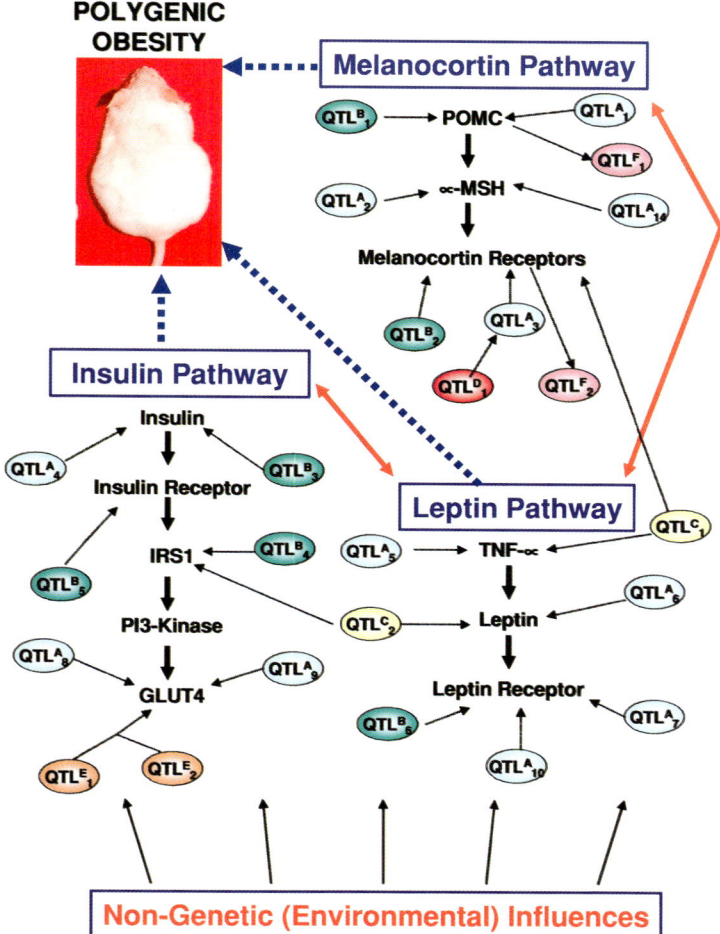

**FIGURE 3.** Hypothesized genetic architecture for obesity predisposition. Simplified and generalized components of example pathways (out of many that combine to regulate obesity) are illustrated. We hypothesize that each of these pathways consists of a complex regulatory cascade with the coordinated and interactive expression of genes playing significant roles in the physiology of the pathway, under the regulation of predisposition genes (quantitative trait loci; QTL) and non-genetic (environmental) influences. In other words, key genes in physiological pathways would for the most part not possess relevant heritable variation; such variation would instead reside within loci that regulate expression levels of these key physiological genes. There could be several potential modes of regulation and complexity for QTL that control expression or activity of physiological genes. QTL$^A$ (blue): QTL can regulate transcription of a physiological gene. For example, a transcription factor within which genetic variation leads to variability in activity or expression levels. QTL$^B$ (green): QTL can regulate post-translational modification of the physiological gene. For example, a kinase or a phosphatase within which genetic variation alters activity and subsequently changes the activity of the physiological gene. QTL$^C$ (yellow): QTL can be pleiotropic factors that regulate multiple physiological genes. QTL$^D$ (red) and QTL$^E$ (orange): QTL can act epistatically to regulate other QTL. For example, certain genetic variation within a QTL is required in order for variation at a second QTL to exert an influence on the expression or activity of a physiological gene (QTL$^D$), or effects of multiple QTL must combine in order to regulate a physiological gene (QTL$^E$). QTL$^F$ (pink): cases where heritable genetic variation does exist directly in the physiological genes, either within regulatory or coding regions, contributing to phenotypic variability in expression or activity of that gene (e.g. *POMC, MC4R*).

## 4.2. Future Directions for Transcriptome Mapping

No published study to date has evaluated multiple tissues in transcriptome mapping. We expect that evaluation of highly relevant yet diverse tissue types will expose different sets of transcript QTLs, even when analyzing the same transcript populations. Discovery of similarities (or differences) across tissues will add power to experimental findings, providing validation especially for cis-acting QTL, and revealing important underlying biology for the traits of interest.

Extension of the paradigm of transcriptome mapping to the proteome and metabolome would also be unique. Although experimental issues (i.e., lack of genome-wide reagents) would render such extension to initially be on a limited scale, it would represent an important test to determine if QTL profiles underlying protein levels parallel those controlling transcription. The question of whether microarrays are valid indicators of actual protein levels (and hence biological activity) is still of major importance, and such results would test and extend this question in regard to comparison of the underlying polygenic control at each step of the central dogma of biology. In regard to metabolomics, discovery of QTL controlling carbohydrate, glucose, lipid and fatty acid metabolism would have particular relevance to obesity research. Our preliminary efforts in this regard have revealed a large number of QTL underlying *de novo* fatty acid synthesis (Allan, 2003; Allan and Pomp, unpublished data).

## 5. STATISTICAL ISSUES

Although transcriptome mapping does not present the need for development of new statistical paradigms relative to traditional transcriptome analysis and QTL mapping, several sophisticated analyses will be required to extract full value from the enormous amount of collected data, and gain valuable insight into genetic control of gene expression. As recently noted by Ariel Darvasi (2003), "I expect that the combining of genetic information and gene expression will hasten the day when genomics delivers on its promise to improve health care. But we must continue striving to develop and apply sophisticated analytical tools for interpreting the vast, complex data sets that are being produced with modern genomic technologies."

Traditionally, these would include analysis of sex-interaction in genetic control of the transcriptome, determination of the role of genomic imprinting in control of genome-wide gene expression, and evaluation of within- and between-founder line genetic variance. Perhaps more importantly, transcriptome mapping represents an extremely challenging scenario for thorough implementation of multiple trait analyses. Initially, this may be best implemented for specific situations, such as genes that are part of the same known pathway, and genes measured in different tissues. Also, when single-trait expression QTL appear to map to the same region, multi-trait analyses can be used to improve precision and significance.

Most QTL analyses have ignored the potential role of gene interactions in the control of trait variation. However, there is mounting evidence that analyses specifically testing for epistasis can both identify QTL that are not otherwise found and explain a greater proportion of the genetic variation (Shimomura et al., 2001; Leamy et al., 2002; Carlborg et al., 2003). In the context of understanding the genetic control of the transcriptome and proteome, it is critical that we extend analyses to include epistasis as this is likely to play an important role in interpretation of the network of gene interactions that contribute to obesity.

When transcriptome mapping is implemented within a very large structured pedigree, the opportunity exists to merge traditional quantitative genetic analyses, such as genetic parameter estimation, with QTL analysis. This would enable estimation of genetic correlations among, and heritabilities of, the sub- (e.g., transcriptional, proteomic, endocrine) and end-point phenotypic traits. As with multi-trait analysis, data reduction is likely required to make this effort feasible and enable extraction of meaningful information. For example, heritabilities can be measured for all endpoint phenotypes, and for all sub-phenotypes for which at least one significant QTL is identified. Genetic correlations can then be estimated for the following sets of traits: A) between each pair of endpoint phenotypes; B) between each sub-phenotype for which heritability is estimated significantly different from zero; and between traits in categories A and B.

Genetic parameter estimation will add unique value to this research paradigm. A strong heritability of a sub-phenotype should coincide with presence of strong evidence for QTL, providing validation of the process. More importantly, genetic correlation between a sub-phenotype and body fat levels will be critical to differentiate among, and rank, multiple linked transcriptional and/or proteomic QTLs that could represent positional candidate genes for obesity predisposition. Finally, it is interesting to speculate that genetic correlation analysis can be a useful method for clustering of array results and identification of biologically relevant pathways. This clustering would be an extension of the stratification of obesity phenotypic classes based on combined expression phenotypes (Schadt et al., 2003a) and the use of principal components analysis recently proposed by Lan et al. (2003).

Both QTL analysis and genetic parameter estimation, in the context of transcriptome mapping, have immense computational requirements. It is likely that such efforts, when carried out on a large scale, will require recoding of existing programs to run on supercomputers.

## 6. IMPLEMENTATION

Analysis of complex trait genetic architecture using "Quantitative Genomics", manifested through the transcriptome (and/or proteome and metabolome) mapping paradigm, can be applied to essentially any QTL mapping experiment where samples with spatial and temporal biological relevance have been stored in an appropriate manner. In regard to obesity, we are applying this approach using the polygenic obese M16 line of mice (Hanrahan et al., 1973; Eisen, 1986) and its non-obese ICR control line (Figure 4). Having identified differences between the lines for a wide variety of traits including transcriptional, proteomic and metabolomic phenotypes

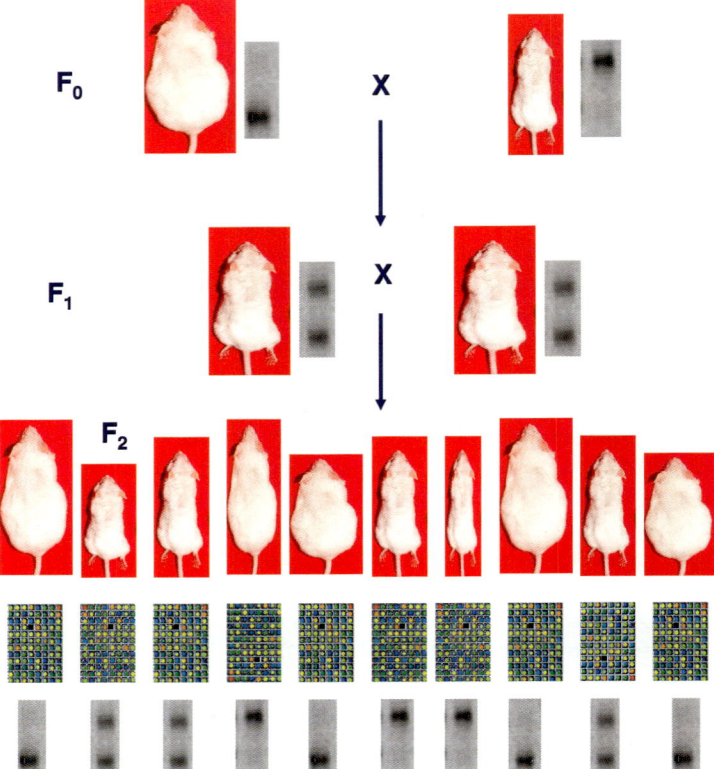

**FIGURE 4.** Application of the transcriptome mapping approach to body weight and obesity using an F2 cross between a line selected for rapid weight gain (M16, top left) and its unselected control line (ICR, top right). Individuals in the segregating F2 population (images here are graphical representations created from a picture of a single mouse) will be phenotyped for body weight, body composition and feed intake. Additionally, tissue samples from F2 individuals will be assayed using a microarray containing most expressed murine genes. Alternative parental line forms of DNA markers will segregate and can be tracked in the F2 population, facilitating a QTL analysis for both weight and composition end-point phenotypes and gene expression sub-phenotypes.

with relevance to energy balance (Pomp et al., 2002; Allan, 2003), we have established a large $F_2$ QTL mapping population and have phenotyped all $\sim$1,200 mice for growth, fatness and feed consumption. Tissues with relevance to energy balance have been stored for sub-phenotyping, while DNA has been extracted and an initial panel of 80 genome-wide informative microsatellites has been genotyped. Furthermore, the $F_2$ was designed to provide for large $3/4$- and $1/2$-sib families to enable genetic parameter estimation for all phenotypes measured across the population. By applying large-scale transcriptional (e.g., Affymetrix, Santa Clara, CA), proteomic (e.g., BD Biosciences (BD Powerblot), San Diego, CA) and metabolomic (e.g., Lipomics Technologies, West Sacramento, CA) phenotyping to the M16 x

ICR $F_2$ QTL mapping population, our goal is to advance the positional cloning of obesity polygenes and begin to understand the genetic architecture of obesity predisposition.

A much broader and extremely powerful platform would be provided by development of a large cohort of recombinant inbred lines (RIL) developed from a multi-way cross of strains representing the majority of phenotypic diversity available in mice (Williams et al., 2002a). A set of 1000 RIL originating from a cross of 8 inbred lines (see Vogel (2003)) would theoretically achieve 0.1 cM precision ($\sim$100,000 bp) when mapping QTL with additive effects of $> 0.25$ SD. Applying transcriptome mapping to this organized assortment of defined recombinational breakpoints would dramatically increase the rate of positional cloning of genes underlying QTL, and would significantly enhance the understanding of the genetic architecture of complex trait predisposition for a wide variety of agriculturally and biomedically relevant phenotypes.

It is prudent to acknowledge that, despite the potential power and breadth of the transcriptome mapping approach, it has important drawbacks and limitations that can and will restrict its utility. We will use examples from recent and well characterized gene discoveries in livestock species to generally illustrate the limitations of transcriptome mapping, assuming hypothetically that this approach were able to have been applied in each case.

One important issue is that some QTL may not be manifested by changes in steady-state levels of mRNA. Not only would transcriptome mapping fail to identify correct candidate genes underlying such QTL, it may in fact mislead the investigator into examining the wrong candidates. For example, the double muscling phenotype in cattle is known to be caused by mutations in the *myostatin* gene, but these mutations are not manifested by changes in mRNA levels but rather by alterations in protein function (Kambadur et al., 1997). Transcriptome mapping would not have identified *myostatin* as a candidate gene in a resource population segregating the double muscling phenotype, while any other expression QTL falling on the cis-diagonal (see Figure 2), in the chromosomal region where double muscling had been mapped to, may have been falsely identified as candidate genes. The approach would still, however, provide important information on transcriptional changes that are downstream from the QTL's effect and which are important in the context of understanding the overall genetic architecture of the trait.

Since the transcriptome mapping approach relies on gene expression phenotypes, it is critical that selection of both spatial (what tissue) and temporal (when the tissue is collected) coordinates captures as much significant biology as possible. An example of this is clearly demonstrated in the recent finding of a QTL represented by a regulatory mutation in IGF2 causing a major effect on muscle growth in pigs (Van Laere et al., 2003). Given that this mutation is manifested by gene expression changes in postnatal skeletal and cardiac muscle but not in fetal muscle or postnatal liver, transcriptome mapping would have been of immediate assistance in finding this mutation only if postnatal muscle was evaluated. In cases such as obesity, multiple tissues are implicated in control of specific pathways that contribute to the end-phenotype, and expression of many genes will vary over time and across environments (e.g., diet). Thus, a thorough transcriptome mapping effort would constitute a massive undertaking involving multiple tissues, time points, environments and genetic backgrounds that, combined with the high cost of microarrays, is likely beyond the scope of most research budgets.

Given that significant effort and expense are invested in data collection using transcriptome mapping, it is therefore imperative that data and results be made broadly available to the research community. One such powerful and useful environment is provided by WebQTL (Wang et al., 2003a; http://www.webqtl.org/). WebQTL is a web-based package for complex trait analysis, and a tool for multidimensional searches among large data sets derived from high-throughput analysis techniques. Furthermore, exploring these data sets in a systematic way will be a challenge. This challenge has already been partially addressed by those describing molecular biology and genetics of simpler eukaryotes. For example, the GRID is a database for genetic or molecular interactions among products of yeast genes (Breitkreutz et al., 2003). Such tools and others like them will be directly applicable to exploration of obesity-related interactions uncovered in transcriptome mapping.

ACKNOWLEDGEMENTS. This research is a contribution of the University of Nebraska Agricultural Research Division and was supported in part by funds provided through the Hatch Act. The authors are grateful to Dale Van Vleck, Joao Rocha, Kari Elo, Rob Williams, David Threadgill, Ken Manly and Chris Haley for useful discussions. Some research in progress discussed in this paper is part of a fruitful and ongoing collaboration with Dr. Gene Eisen at North Carolina State University. The source of the figures and much of the text in this Chapter is from the *Journal of Animal Science*, and copyright is owned by the American Society of Animal Science. The authors are grateful to the American Society of Animal Science for permission to reproduce this information.

## 7. REFERENCES

Allan, M.F., 2003, Integration of genomic, proteomic and metabolomic analyses towards polygene discovery in mice selected for 3 to 6 week weight gain, PhD Dissertation, University of Nebraska—Lincoln.

Barsh, G.S., and Schwartz, M.W., 2002, Genetic approaches to studying energy balance: perception and Integration, *Nature Reviews Genetics* 3: 589–600.

Bray, G.A., and Tartaglia, L.A., 2000, Medicinal strategies in the treatment of obesity. *Nature* 404: 672–677.

Bray, G., and Bouchard, C., 1997, Genetics of human obesity: research directions, *FASEB J* 11: 937–945.

Breitkreutz, B.J., Stark, C., and Tyers, M., 2003, The GRID: the General Repository for Interaction Datasets, *Genome Biol.* 4:R23

Brem, R.B., Yvert, G., Clinton, R., and Kruglyak, L., 2002, Genetic dissection of transcriptional regulation in budding yeast, *Science* 296: 752–755.

Brockmann, G.A., and Bevova, M.R., 2002, Using mouse models to dissect the genetics of obesity, *Trends in Genetics* 18: 367–376.

Brockmann, G.A., Kratzsch, J., Haley, C.S., Renne, U., and Schwerin, M., 2000, Single QTL effects, epistasis, and pleiotropy account for two-thirds of the phenotypic $F_2$ variance of growth and obesity in DU6i x DBA/2 mice, *Genome Res.* 10: 1941–1957.

Carlborg, Ö., Kerje, S., Schutz, K., Jacobsson, L., Jensen, P., and Andersson, L., 2003, A Global Search Reveals Epistatic Interaction between QTLs for Early Growth in the Chicken, *Genome Res.* 13: 413–421.

Chagnon, Y.C., Rankinen, T., Snyder, E.E., Weisnagel, S.J., Perusse, L., and Bouchard, C., 2003, The human obesity gene map: the 2002 update, *Obes. Res.* 11: 313–367.

Chua, S.C. Jr. 1997, Monogenic models of obesity, *Behav. Genet.* 27: 277–284.

Comuzzie, A.G., and Allison, D.B., 1998, The search for human obesity genes, *Science* 280: 1374–1377.

Consoli, L., Lefevre, A., Zivy, M., de Vienne, D., and Damerval, C., 2002, QTL analysis of proteome and transcriptome variations for dissecting the genetic architecture of complex traits in maize, *Plant Mol. Biol.* 48: 575–581.

Darvasi, A., 2003, Genomics: Gene expression meets genetics, *Nature.* 422: 269–270.

Damerval, C., Maurice, A., Josse, J.M., and de Vienne, D., 1994, Quantitative trait loci underlying gene product variation: a novel perspective for analyzing regulation of genome expression, *Genetics* 137: 289–301.

de Vienne, D., Maurice, A., Josse, J.M., Leonardi, A., and Damerval, C., 1994, Mapping factors controlling genetic expression, *Cell. Mol. Biol.* (Noisy-le-grand) 40: 29–39.

Dietrich, W.F., Miller, J., Steen, R., Merchant, M.A., Damronboles, D., Husain, Z., Dredge, R., Daly, M.J., Ingalls, K.A., Oconnor, T.J., Evans, C.A., Deangelis, M.M., Levinson, D.M., Kruglyak, L., Goodman, N., Copeland, N.G., Jenkins, N.A., Hawkins, T.L., Stein, L., Page, D.C., and Lander, E.S., 1996, A comprehensive genetic map of the mouse genome, *Nature* 380: 149–152.

Doerge, R.W., 2002, Mapping and analysis of quantitative trait loci in experimental populations, *Nat. Rev. Genet.* 3: 43–52.

Dumas, P., Sun, Y., Corbell, G., Trenblay, S., Pausova, Z., Kren, V., Krenova, D., Pravence, M., Hamet, P., and Treblay, J., 2000, Mapping of quantitative trait loci (QTL) of differential stress gene expression in rat recombinant inbred strains, *J. of Hypertension* 18: 545–551.

Eisen, E.J., 1986, Maturing patterns of organ weights in mice selected for rapid postweaning gain, *Theor. Appl. Genet.* 73: 148–157.

Eisen, E.J., 1989, Selection experiments for body composition in mice and rats: a review, *Livest. Prod. Sci.* 23: 17–32.

Elo, K., 2003, Towards integrated genomic approaches—lessons from energy metabolism studies, in: *Proceedings of the NJF's 22ⁿᵈ Congress*, Turku, Finland, pages 222–227.

Esterbauer, H., Schneitler, C., Oberkofler, H., Ebenbichler, C., Paulweber, B., Sandhofer, F., Ladurner, G., Hell, E., Strosberg, A.D., Patsch, J.R., Krempler, F., and Patsch, W., 2001, A common polymorphism in the promoter of *UCP2* is associated with decreased risk of obesity in middle-aged humans, *Nat. Genet.* 28: 178–183.

Farooqi, I.S., Keogh, J.M., Yeo, G.S., Lank, E.J., Cheetham, T., and O'Rahilly, S., 2003, Clinical spectrum of obesity and mutations in the melanocortin 4 receptor gene, *N. Engl. J. Med.* 348: 1085–95.

Festing, M.F.W., 1979, The inheritance of obesity in animal models of obesity, in: *Animal Models of Obesity.* M.F.W. Festing (ed.) pp. 15–37, MacMillan Press, London.

Finkelstein, E.A., Fiebelkorn, I.C., and Wang, G., 2003, National medical spending attributable to overweight and obesity: how much, and who's paying? *Health Aff (Millwood).* Suppl: W3-219–26.

Flegal, K.M., Carroll, M.D., Ogden, C.L., and Johnson. C.L., 2002, Prevalence and trends in obesity among US adults, 1999–2000, *JAMA* 288: 1723–1727.

Garrod, A., 1902, The incidence of alkaptonuria: A study in chemical individuality, *Lancet* 2: 1616–1620.

Haldane, J.B., Sprunt, A.D., and Haldane, N.M., 1915, Reduplication in mice, *Science* 5: 133–135.

Haley, C.S., Knott, S.A., and Elsen, J.M., 1994, Mapping Quantitative Trait loci in crosses between outbred lines using least squares, *Genetics* 136: 1195–1207.

Hanrahan, J.P., Eisen, E.J., and Legates, J.E., 1973, Effects of population size and selection intensity on short-term response to selection for post-weaning gain in mice, *Genetics* 73: 513–530.

Heo, M., Leibel, R.L., Boyer, B.B., Chung, W.K., Koulu, M., Karvonen, M.K., Pesonen, U., Rissanen, A., Laakso, M., Uusitupa, M.I., Chagnon, Y., Bouchard, C., Donohoue, P.A., Burns, T.L., Shuldiner, A.R., Silver, K., Andersen, R.E., Pedersen, O., Echwald, S., Sorensen, T.I., Behn, P., Permutt, M.A., Jacobs, K.B., Elston, R.C., Hoffman, D.J., and Allison, D.B., 2001, Pooling analysis of genetic data: the association of leptin receptor (LEPR) polymorphisms with variables related to human adiposity, *Genetics* 159: 1163–1178.

Hixson, J.E., Almasy, L., Cole, S., Birnbaum, S., Mitchell, B.D., Mahaney, M.C., Stern, M.P., MacCluer, J.W., Blangero, J., and Comuzzie, A.G., 1999, Normal variation in leptin levels is associated with polymorphisms in the proopiomelanocortin gene, *POMC. J. Clin. Endocrinol. & Metab.* 84: 3187–3191.

Hilbert, P., Lindpainter, K., Beckmann, J.S., Serikawa, T., Soubrier, F., Dubay, C., Cartwright, P., De Gouyon, B., Julie, C., Takahasi, S., Vincent, M., Ganten, D., Georges, M., and Lathrop, G.M., 1991, Chromosomal mapping of two genetic loci associated with blood-pressure regulation in hereditary hypertensive rats, *Nature* 353: 521–529.

Hitzemann, R., Malmanger, B., Reed, C., Lawler, M., Hitzemann, B., Coulombe, S., Buck, K., Rademacher, B., Walter, N., Polyakov, Y., Sikela, J., Williams, R.W., Flint, J., and Talbot, C., 2003, A strategy for integration of QTL, gene expression, and sequence analyses, *Mammalian Genome* 14: 733–747.

Jacob, H.J., Lindpainter, K., Lincoln, S.E., Kusumi, K., Bunker, R.K., Mao, Y.-P., Ganten, D., Dzau, V.J., and Lander, E.S., 1991, Genetic mapping of a gene causing hypertension in the stroke-prone rat, *Cell* 67: 213–224.

Jansen, R.C., and Nap, J.P., 2001, Genetical genomics: the added value from segregation, *Trends Genet.* 17: 388–391.

Jansen, R.C., 2003, Studying complex biological systems using multifactorial perturbation, *Nature Reviews Genetics* 4: 145–151.

Kambadur R., Sharma, M., Smith, T.P., and Bass, J.J., 1997, Mutations in myostatin (GDF8) in double-muscled Belgian Blue and Piedmontese cattle, *Genome Res.* 7: 910–6.

Korstanje, R., and Paigen, B., 2002, From QTL to gene: the harvest begins, *Nat. Genet.* 31: 235–236.

Koza, R.A., Hohmann, S.M., Guerra, C., Rossmeisl, M., and Kozak, L.P., 2000, Synergistic gene interactions control the induction of the mitochondrial uncoupling protein (Ucp1) gene in white fat tissue, *J. Biol. Chem.* 275: 34486–34492.

Lan, H., Stoehr, J.P., Nadler, S.T., Schueler, K.L., Yandell, B.S., and Attie, A.D., 2003, Dimension reduction for mapping mRNA abundance as quantitative traits, *Genetics* 164: 1607–14.

Lander, E.S., and Botstein, D., 1989. Mapping Mendelian factors underlying quantitative traits using RFLP linkage maps. Genetics 121: 185–199.

Leamy, L.J., Routman, E.J., and Cheverud, J.M., 2002, An epistatic genetic basis for fluctuating asymmetry of mandible size in mice, *Evolution Int. J. Org. Evolution* 56: 642–653.

Lembertas, A.V., Perusse, L., Chagnon, Y.C., Fisler, J.S., and Warden, C.H., 1997, Identification of an obesity quantitative trait locus on mouse chromosome 2 and evidence of linkage to body fat and insulin on the human homologous region 20q, *J. Clin. Invest.* 100: 1240–1247.

Martin, B., Nienhuis, J., King, G., and Schaefer, A., 1989, Restriction fragment length polymorphisms associated with water use efficiency in tomato, *Science* 243: 1725–1728.

McPherron, A.C., and Lee, S.J., 1996, The transforming growth factor b superfamily, in: *Growth Factors and Cytokines in Health and Disease*, D. LeRoith and C. Bondy, Eds., pp. 357–393. JAI Press, Greenwich, CT.

Mehrabian, M., Wen, P.Z., Fisler, J., Davis, R.C., and Lusis, A.J., 1998, Genetic loci controlling body fat, lipoprotein metabolism, and insulin levels in a multifactorial mouse model, *J. Clin. Invest.* 101: 2485–2496.

Michuad, E.J., Bultman, S.J., Klebig, M.L., van Vugt, M.J., Stubbs, L.S., Russell, L.B., and Woychik, R.P., 1994, A molecular model for the genetic and phenotypic characteristics of the mouse lethal yellow ($A^y$) mutation, *Proc. Natl. Acad. Sci.* 91: 2562–2566.

Montooth, K.L., Marden, J.H., and Clark, A.G., 2003, Mapping determinants of variation in energy metabolism, respiration and flight in Drosophila, *Genetics* 165: 623–35.

Nadeau, J.H., and Frankel, W.N., 2000, The roads from phenotypic variation to gene discovery: mutagenesis versus QTLs, *Nat. Genet.* 25: 381–384.

Oliver, B., Parisi, M., and Clark, D., 2002, Gene expression neighborhoods, *J. Biol.* 1: 4.

Paterson, A.H., Lander, E.S., Hewitt, J.D., Peterson, S., Lincoln, S.E., and Tanksley, S.D., 1988, Resolution of quantitative traits into Mendelian factors by using a complete linkage map of restriction fragment length polymorphisms, *Nature* 335: 721–726.

Pomp, D., 1997, Genetic dissection of obesity in polygenic animal models, *Behav. Genet.* 27: 285–306.

Pomp, D., 1999, Animal models of obesity, *Molecular Medicine Today* 5: 459–460.

Pomp, D., Jerez, N., Allan, M.F., and Eisen, E.J., 2002, Integrated genomic, proteomic and metabolomic dissection of polygenic selection response for murine growth and fatness, in: *Proceedings of the 7th World Congress on Genetics Applied to Livestock Production*, Montpellier, France.

Rich, S.S., 1990, Mapping genes in diabetes, *Diabetes* 39: 1315–1319.

Rocha, J.L., Van Vleck, L.D., Eisen, E.J., and Pomp, D., 2004a, A large-sample QTL study in mice: I. Growth. *Mamm. Genome* (In Press)

Rocha, J.L., Van Vleck, L.D., Eisen, E.J., and Pomp, D., 2004b, A large-sample QTL study in mice: II. Organ and body composition traits, *Mamm. Genome* (In Press)

Rohrer, G.A., Wise, T.H., Lunstra, D.D., and Ford, J.J., 2001, Identification of genomic regions controlling plasma FSH concentrations in Meishan-White Composite boars, *Physiol. Genomics* 6: 145–151.

Rosen, C.J., Churchill, G.A., Donahue, L.R., Shultz, K.L., Burgess, J.K., Powell, D.R., Ackert, C., and Beamer, W.G., 2000, Mapping quantitative trait loci for serum insulin-like growth factor-1 levels in mice, *Bone* 27: 521–528.

Sax, K., 1923, The association of size differences with seed-coat pattern and pigmentation in Phaseolusvulgaris, *Genetics* 8: 552–560.

Schadt, E.E., Monks, S.A., and Friend, S.H., 2003, A new paradigm for drug discovery: integrating clinical, genetic, genomic and molecular phenotype data to identify drug targets. *Biochem. Soc. Trans.* 31: 437–443.

Schadt, E.E., Monks, S.A., Drake, T.A., Lusis, A.J., Che, N., Colinayo, V., Ruff, T.G., Milligan, S.B., Lamb, J.R., Cavet, G., Linsley, P.S., Mao, M., Stoughton, R.B., and Friend, S.H., 2003, Genetics of gene expression surveyed in maize, mouse and man, *Nature* 422: 397–402.

Schwartz, M.W., Woods, S.C., Seeley, R.J., Barsh, G.S., Baskin, D.G., and Leibel, R.L., 2003, Is the energy homeostasis system inherently biased toward weight gain? *Diabetes* 52: 232–238.

Shimomura, K., Low-Zeddues, S.S., King, D.P., Steeves, T.D.L., Whiteley, A., Kushla, J., Zemenides, P.D., Lin, A., Vitaterna, M.H., Churchill, G.A., and Takahashi, J.S., 2001, Genome-wide epistatic interaction analysis reveals complex genetic determinants of circadian behavior in mice, *Genome Research* 11: 959–980.

Spellman, P.T., and Rubin, G.M., 2002, Evidence for large domains of similarly expressed genes in the Drosophila genome, *J. Biol.* 1: 5.

Sturtevant, A.H., 1913, The linear arrangement of six sex-linked factors in Drosophila, as shown by their mode of association, *J. Exp. Zool.* 14: 43–59.

Van Laere A.S., Nguyen, M., Braunschweig, M., Nezer, C., Collette, C., Moreau, L., Archibald, A.L., Haley, C.S., Buys, N., Tally, M., Andersson, G., Georges, M., and Andersson, L.A., 2003, A regulatory mutation in IGF2 causes a major QTL effect on muscle growth in the pig, *Nature* 425: 832–6.

Vogel, G., 2003, Scientists dream of 1001 complex mice, *Science* 301: 456–457.

Wang, J., Williams, R.W., Chesler, E.J., and Manly, K.F., 2003a, WebQTL: Web-based complex trait analysis, *Neuroinformatics* 1: 299–308.

Wang, Y.X., Lee, C.H., Tiep, S., Yu, R.T., Ham, J., Kang, H., and Evans, R.M., 2003b, Peroxisome-Proliferator-Activated Receptor delta Activates Fat Metabolism to Prevent Obesity, *Cell* 113: 159–170.

Wayne, M.L., and McIntyre, L.M., 2002, Combining mapping and arraying: An approach to candidate gene identification, *Proc. Natl. Acad. Sci.* 99: 14903–14906.

Wesolowski, S.R., Allan, M.F., Nielsen, M.K., and Pomp, D., 2003, Evaluation of hypothalamic gene expression in mice divergently selected for heat loss, *Physiol. Genomics* 13: 129–137.

Williams, R.W., Broman, K.W., Cheverud, J.M., Churchill, G.A., Hitzemann, R.W., Hunter, K.W., Mountz, J.D., Pomp, D., Reeves, R.H., Schalkwyk, L.C., and Threadgill, D.W., 2002a, A collaborative cross for high-precision complex trait analysis, in: *1ˢᵗ Workshop Report of the Complex Trait Consortium*, September, 2002. www.complextrait.org/Workshop1.pdf

Williams, R.W., Shou, S., Lu, L., Qu, Y., Wang, J., Manly, K.F., Chesler, E.J., Hsu, H.C., Mountz, J.D., and Threadgill, D.W., 2002b, Genomic Analysis of Transcriptional Networks: Combining Microarrays with Complex Trait Analysis, in: *Proceedings of the 1st Annual Complex Trait Consortium Meeting*, Memphis, Tennessee. http://www.complextrait.org/ctc2002/williams.html

Yvert, G., Brem, R.B., Whittle, J., Akey, J.M., Foss, E., Smith, E.N., Mackelprang, R., and Kruglyak, L., 2003, Trans-acting regulatory variation in *Saccharomyces cerevisiae* and the role of transcription factors, *Nat. Genet.* 35: 57–64.

Zeng, Z.-B., 1993, Theoretical basis for separation of multiple linked gene effects in mapping quantitative trait loci, *Proc. Natl. Acad. Sci.* 90: 10972–10976.

Zeng, Z.-B. 1994, Precision mapping of quantitative trait loci. Genetics 136: 1457–1468.

Zhang, Y., Proenca, R., Maffei, M., Barone, M., Leopold, L., and Friedman, J., 1994, Positional cloning of the mouse obese gene and its human homologue, *Nature* 372: 425–432.

## Chapter 12

# Mining the EST Databases to Determine Evolutionary Events in the Legumes and Grasses

Jessica A. Schlueter, Phillip Dixon, Cheryl Granger, and Randy C. Shoemaker

## 1. INTRODUCTION

Gene and genome duplication have long been accepted as a driving force in the evolution and expansion of eukaryotic genomes (Stebbins, 1950; Ohno, 1970). These phenomena, whole genome/polyploidy events, are seen across numerous eukaryotic species and are highly prevalent within the plant kingdom. Gene duplication is seen as a mechanism for the creation of genetic diversity, genome expansion, and creation of new gene functions. It also often leads to silenced genes or pseudogenes (Pickett and Meeks-Wagner, 1995).

Duplication can occur by a variety of mechanisms: duplication of regions or segments of chromosomes, tandem duplication, dispersed duplication, reverse transcriptase mediated DNA insertion from RNA intermediates, and whole genome duplication or polyploidy. After a polyploidy event, duplicated regions begin to diverge from one another at both the sequence and chromosomal levels (Pickett and Meeks-Wagener, 1995), a process referred to as 'diploidization' (Stebbins, 1966). This is a concept originally envisioned by Clausen in 1941.

**Jessica A. Schlueter, Phillip Dixon, Cheryl Granger, and Randy C. Shoemaker**    USDA-ARS-CICGR Unit, Agronomy Hall, Iowa State University, Ames, Iowa 50011

*Genome Exploitation: Data Mining the Genome*, edited by J. Perry Gustafson, Randy Shoemaker, and John W. Snape.
Springer Science + Business Media, New York, 2005.

'Diploidization' is caused by additions, deletions, mutations, and rearrange-ments that rapidly inhibit non-homologous pairing of [2]chromosomal tetravalents (Ohno, 1970). In addition, duplicated genes may undergo a change in function or silencing of one of the copies (Pickett and Meeks-Wagner, 1995). Despite these numerous genomic changes, it is often possible to detect homoeologous chromosomal regions from ancient duplication events in diploid species.

Through the analysis of duplicated genes it may be possible to gain an understanding of the mechanisms and processes of diploidization and genome evolution. For example, insights into the evolutionary history of Arabidopsis have been brought to light through studies focusing on duplicated genes and genome segments. In 1994, Kowalski et al. found single copy markers in *Brassica oleracea* mapped to duplicate regions in the Arabidopsis genome. Comparative genomics between Arabidopsis and soybean and Arabidopsis and tomato showed strong evidence for segmental and possible whole genome duplication in Arabidopsis (Grant et al., 2000; Ku et al., 2000). Blanc et al. (2000) suggested that Arabidopsis could actually be a degenerate tetraploid. Analysis of the whole Arabidopsis genome sequence has revealed that more than one large-scale genome duplication may have occurred, although the exact number of rounds is still debated (The Arabidopsis Genome Initiative, 2000; Vision et al., 2000; Simillion et al., 2002; Blanc et al., 2003).

As data from the rice genome sequencing projects have been compiled, it has become apparent that rice also has extensive gene duplication, possibly the result of a massive genome duplication event (Yu et al, 2002). The general belief in the existence of a simple diploid plant species is probably a false one, with most all plant genomes being the result of extensive duplication and reshuffling.

## 2. EVOLUTIONARY HISTORY OF LEGUMES AND GRASSES

Monocot/dicot divergence is thought to have occurred approximately 200 MYA (Wolfe et al., 1989, Goff et al. 2002). Despite the breadth of genetic and genomic research being conducted on many key plants, e.g., soybean (*Glycine max* L.), maize (*Zea Mays* L.) and, rice (*Oryza sativa* L.), and barrell medic, the evolutionary history of their genomes remains relatively unresolved. While whole genome sequence will provide a fuller picture of the history of these genomes, the ongoing EST projects provide an excellent resource to detect genome events.

Considerable differences exist in the size of these plant genomes. The size of the genome is independent of chromosome number. The soybean genome (2n = 40) is comprised of about 1.1 Mbp/C, making it about twice the size of the Medicago (2n = 16) genome (Arumganthanan and Earle, 1991). The maize genome (2n = 20) is approximately six times larger than the genome of rice (2n = 24) and approximately twice the size of the soybean genome (Arumganthanan and Earle, 1991).

Most genera of the Phaseolae have a genome complement of 2n = 22, suggesting that soybean may have been derived from a diploid ancestor (n = 11) which underwent aneuploid loss to n = 10 and subsequent tetraploidization followed by diploidization (Lackey, 1980). More than 90% of soybean RFLP probes detected more than two fragments, further suggesting that large amounts of the soybean genome may have undergone some form of genome duplication (Shoemaker et al., 1996). Combined data from nine different mapping populations uncovered extensive homoeologous relationships among linkage groups (Shoemaker et al., 1996). The detection of 'nested' duplications suggested that at least one of the original genomes of soybean might have been duplicated prior to the major tetraploidization event (Shoemaker et al., 1996; Lee et al., 2001).

Both soybean and barrel medic belong to the large flowering family Fabaceae and share the distinguishing ability to symbiotically fix nitrogen. The evolutionary history of Medicago is not as well studied as that of soybean. There is, however, evidence of gene duplication in Medicago; glutathione synthetase exists in two copies, most likely through a tandem duplication (Frendo et al., 2001). Additionally, Medicago chromosome 5 exhibits orthology with another region on the same chromosome showing evidence of segmental duplication (Gualtieri and Bisseling, 2002). It has been suggested that Medicago has a simple genome structure, with only tandem and segmental duplications (Kulikova et al., 2001). However, prior to genome sequencing, the Arabidopsis genome was considered a true diploid with only 15% of the loci being duplicated (McGrath et al., 1993). Now, whole genome duplications are well accepted as a driving force in the evolution of the Arabidopsis genome (The Arabidopsis Genome Initiative 2000; Vision et al., 2000; Simillion et al., 2002; Blanc et al., 2003). As the genome sequence of Medicago becomes available, it is possible that the same will be true as with Arabidopsis. Although extensive regions of synteny may not be commonplace among legumes, microsynteny may be observed frequently. Yan et al. (in press) estimates that only about half of 50 soybean contigs evaluated showed some degree of microsynteny with Medicago.

The grasses have had more extensive comparative genetic studies than the legumes. In 1993, Ahn and Tanksley showed homoeologous relationships between rice and maize through comparative genetic maps. Further, microcolinearity in markers is observed in the grasses (Devos and Gale, 1997). This colinearity is conserved over 60 million years and allows the combination of numerous genetic maps into one grass map, a tool for comparisons of homoeologous regions (Moore et al., 1995).

Evidence of duplicated genes in maize has been documented as early as 1951 (Rhoades). This has been further supported by multiple-copy RFLP's showing conserved order on homeologous chromosomes (Helentjaris et al., 1988). More than 72% of the loci in maize are estimated to exist in duplicate (Ahn and Tanksley, 1993). Further, the results of Ahn and Tanksley showed that major chromosomal rearrangements have occurred as the genome rediploidized. Helentjaris et al. (1988) suggested that maize is an allopolyploid with an ancestral haploid chromosome number of five. Maize has also been proposed to be a segmental allotetraploid, with divergence of the diploid progenitors approximately 20 million years ago with a tetraploid event 11 to 16 million years ago (Gaut and Dobley, 1997; Gaut et al., 2000).

The rice genome, however, has been considered a simple diploid, similar to Medicago. A molecular mapping study in rice found that only 20% of the markers were multiple-copy (McCouch et al., 1988). The few that were identified in more than one copy were attributed to transposon activity (McCouch et al., 1988). However, there is now extensive evidence of a chromosomal duplication between chromosomes 11 and 12 of the rice genome (Nagamura et al., 1995). Both of the rice genome sequencing groups have identified multiple gene duplications in the rice genome (Yu et al., 2002; Goff et al., 2002). Further, Salse et al. (2002) has identified duplicated rice sequence that exists in only one copy in Arabidopsis suggesting rice has chromosomal duplications.

Recent genome projects have generated an unprecedented number of genic sequences for soybean (Shoemaker et al., 2002), barrel medic (Gyorgyey et al., 2000), maize (Lunde et al., 2003), and rice (Ewing et al., 1999). The sequences generated by these projects represent a largely untapped resource for dissection of gene families, construction of phylogenetic trees and studies of evolutionary histories of genomes.

## 3. IDENTIFICATION OF DUPLICATED CONTIGS

Tentative contigs for *Glycine max, Medicago truncatula*, maize, and rice were obtained from the TIGR genome database, available at http://www.tigr.org/tdb/agi/. Each tentative contig (TC) is comprised of several EST's that are related to one another by contiguous overlapping arrangements. A TC is subject to change as new EST's are added to the database and TC's are re-established. For this analysis the sets assembled on May 13, 2002 were used in all analyses.

Each species-specific TC set was 'piped' through the program Orffinder, part of the EMBOSS bioinformatics package (Rice et al., 2000). Once all possible open reading frames (ORF's) were obtained, a perl script was used to obtain the longest orf for each TC. These species-specific long orf TC's were used in a similarity search against themselves using a tblastx search algorithm (Altschul, 1990). Parsing was done to identify potential paralogs or pairs using the program MuSeqBox (Xing and Brendel, 2000) to tabularize the BLAST output and then a perl script. The perl script identified potential paralogs with a query and subject coverage of greater than or equal to 80% and by having only the query and the subject fit the criteria. Identified paralogs were required to be reciprocal; if *a* identified *b*, then *b* must identify *a*. The criteria used identified pairs that are doubles and not triples, etc.

Potentially paralogous TCs comprised of ESTs of mixed genotype were re-contiged with CAP3 (Huang and Madan, 1999) with a 40bp overlap and 95% similarity requirement. For *G. max*, the acceptable genotype was a combination of Williams and Williams82; for *M. truncatula*, the genotype Jemalong; for *Z. mays*, the genotype B73; and for *O. sativa*, the genotype Nipponbare. Williams and Williams82 are near-isogenic lines and should not introduce genotype specific polymorphisms except for the introgressed region. If any contig contained two or fewer EST's, the contig and its associated pair were removed from the data set.

These recontigged TC consensus sequences plus TIGR TC consensus sequences, used for contigs that were not of mixed genotype, comprised the complete

**Table 1**
**Paralogs Identified Through Analyses of EST Collections.**

| | TIGR TCs | Singletons[a] | Total ESTs | Paralogs Identified | Percent Genotype[b] | Genotype |
|---|---|---|---|---|---|---|
| *Glycine max* | 20,642 | 29,039 | 236,461 | 256 | 55% | Williams/ Williams82 |
| *Medicago truncatula* | 16,086 | 17,658 | 164,304 | 291 | 91% | Jemalong |
| *Zea mays* | 17,465 | 12,930 | 148,429 | 95 | 32% | B73 |
| *Oryza sativa* | 13,745 | 18,456 | 103,340 | 50 | 86% | Nipponbare |

[a]Singletons are the number of single ESTs not placed in a TIGR TC.
[b]Percent genotype reflects the percent of the genotype in the total ESTs.

set of genotype specific potential paralogous TCs (gpTCs). Each gpTC was run through the program Orffinder and the abovementioned perl script to obtain the longest putative open reading frame.

Each gpTC longest orf was imported into MegAlign, a sequence alignment program part of the Lasergene package (DNASTAR, Inc). Alignments were done between pairs using the Wilbur-Lipman method with default parameters on virtural translations of the potential open reading frames. Once a common translated (protein) sequence was found, the alignment was back translated to DNA, cropped to between the start and stop codons, and saved in GCG MSF format. If the full sequence between start and stop codons was not available, then the region in each gpTC showing full overlap was used. If the open reading frame was not found with the longest orf obtained above, a blastx was performed with the original gpTC consensus sequence against the NCBI nonredundant database to obtain the most likely open reading frame. That open reading frame was used for the virtual translation in MegAlign and analysis proceeded as above.

Over 250 potential duplicated genes were identified for both soybean and Medicago (Table 1). Surprisingly, for maize, a known ancient polyploid, only 95 contigs were identified as potential duplicates. This might be an artifact of the BLAST stringency, but is most likely due to the limitation placed on genotype; only 32% of the maize EST's were B73. The low number of potential paralogs (50) identified for rice may not be due solely to genotype limitation but instead may be due to the limited number of EST's available for this study. Further, this low number may reflect a genome that has experienced segmental duplications rather than whole genome events (Table 1).

## 4. PAIRWISE DISTANCE MEASURES AND MIXTURES OF NORMAL DISTRIBUTIONS

Pairwise distance measures were obtained by the program Diverge, part of the GCG package (Wisconsin Package). Diverge calculates synonymous and nonsynonymous distances using the methods of Li et al. (1985) as modified by Li (1993) and Pamilo and Bianchi (1993). This method was chosen because it

## Table 2
### Characteristics of Paralogs and Distance Estimates.

|                                       | G. max            | M. truncatuala    | Z. mays           | O. sativa         |
| ------------------------------------- | ----------------- | ----------------- | ----------------- | ----------------- |
| Average Paralog Length                | 474 bp            | 618 bp            | 379 bp            | 380 bp            |
| Average Protein Identity[a]           | 85%               | 82%               | 88%               | 88%               |
| Average Protein Similarity[a]         | 92%               | 90%               | 93%               | 93%               |
| Average Nucleotide Identity[a]        | 82%               | 80%               | 86%               | 82%               |
| Synonymous Distance Range[b]          | $0.007 \pm 0.010$ to $2.59 \pm 1.648$ | $0.007 \pm 0.007$ to $2.88 \pm 1.701$ | $0.007 \pm 0.007$ to $2.39 \pm 1.717$ | $0.012 \pm 0.012$ to $2.91 \pm 3.226$ |

[a] Protein identity and similarity was determined by virtual translation of paralogous pair and alignment by the program Water, part of the GCG package (Wisconsin Package). Nucleotide alignment was performed by the program Water to determined nucleotide identity.
[b] Synonymous distance range is the smallest Ks standard $\pm$ deviation to the largest Ks standard $\pm$ deviation.

is based upon Kimura's (1980) method for distinguishing between transitions and transversions, thereby allowing for unequal rates of nucleotide substitutions. Diverge provides an estimate of the number of synonymous nucleotide substitutions per synonymous site (Ks), nonsynonymous nucleotide substitutions per nonsynonymous site (Kn), as well as the standard deviations for both the Ks and Kn. Table 2 shows characteristics of the paralogs used to determine the Ks and Kn distance measures as well as the range of synonymous distances obtained.

As seen in Table 2, a wide range in standard deviations (sd) was observed that increased as synonymous distances increased. To reduce this effect, both the Ks and sd of Ks were natural-log transformed to allow the sd to be similar across all Ks. Ks values less than 0.05 were removed as done by Kondrashov et al. (2002).

It is assumed that major genome events can be represented as a normal distribution of synonymous distances, and multiple events as a mixture of normal distributions. To determine the most likely number of distributions, or components, of synonymous distances, maximum likelihood statistics were utilized.

A module was written for the statistical software Splus (Mathsoft) to calculate the most probable mixture of distributions that fit the data ranging from one to five components. For each number of potential components, 100 replications were done to determine the most probable fit and to calculate a log likelihood statistic. Two times the change in log likelihood from a simpler model to a more complex model follows a Chi-square distribution with a range of 4 to 6 degrees of freedom and allows the determination of a p-value. Statistical significance was set at $p < 0.05$. Each normal distribution of synonymous distance is comprised of clusters of gene pairs that are assumed to correspond to a large genome event, i.e., duplication.

Across all species, there is evidence for more than one normal distribution within the data. The legumes soybean and Medicago both support a trimodal distribution with high statistical significance (p = 0.005 and p = 0.025, respectively). Figure 1 shows the corresponding normal distributions for soybean and Medicago. The mean for each distribution is the mean of the synonymous distances contained within that distribution. While both soybean and Medicago fit trimodal distributions, the corresponding means are not directly shared between the two. There is no statistical support for a four-component model in either species. Unlike the legumes, the grasses do not share the same number of distributions. Maize supports a trimodal distribution (p = 0.05), and rice a bimodal distribution (p = 0.005). The resulting distributions as well as the means for each are shown in Figure 3.

As saturation is reached in the synonymous distance estimates for all species, at approximately 1.0, the predicted distributions become broader and encompass a greater range of synonymous distances. If we assume that the synonymous distance measures represent the rate of nucleotide substitution, each normal distribution is representative of gene pairs evolving at a similar rate. Relative dates of genome events can be determine by utilizing the means of these normal distributions.

## 5. COALESENCE ESTIMATES

Assuming that the average synonymous substitution rate is linear relative to time, and that the rate is equal across all paralogs, then the relative ages for each observed event can be determined. An estimated synonymous substitution rate of $6.1 \times 10^{-9}$ substitutions per synonymous site per year (Lynch and Conery, 2000), and $6.5 \times 10^{-9}$ substitutions per synonymous site per year (Gaut et al., 1996) were used to determine genome events in the legumes and grasses, respectively. These rates were assumed to be representative of the synonymous substitution rates of our identified paralogs (Gaut and Doebley, 1997).

Under these assumptions, the three statistically significant normal distributions, or genomic events, in soybean are estimated to have occurred approximately 15, 40, and 104 million years ago (MYA) (Figure 1). The three significant distributions observed in Medicago correspond to genomic events occurring at 26, 51, and 117 MYA (Figure 1).

If the distributions observed in both soybean and Medicago are overlaid (Figure 2), events that occurred in concert should have very similar distributions. Doyle and Luckow (2003) estimated that soybean and Medicago had already diverged between 50 and 60 MYA. Therefore, events predating the divergence of the two genera should coincide. Very probably, the 104–117 MYA coalescence times represent a single event that occurred while the two genera shared a common lineage. Therefore, the two genera diverged after that date.

The detection of large, relatively well-maintained duplicated segments in the soybean linkage map (Shoemaker et al., 1996) suggests that the youngest

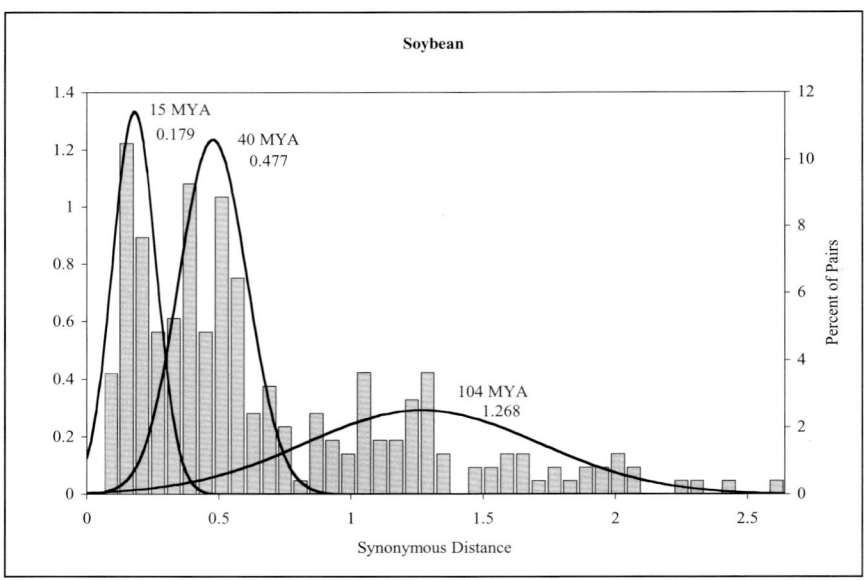

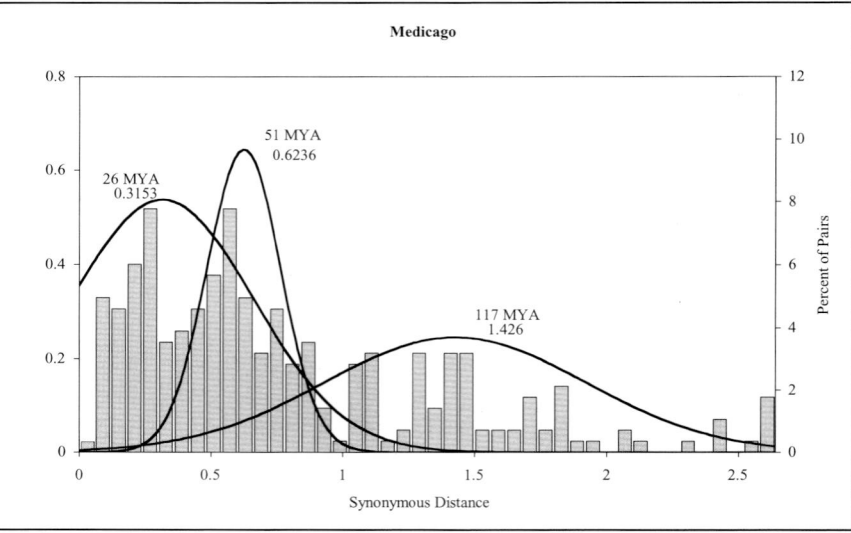

**FIGURE 1.** Histogram of the percent of identified pairs for both soybean and Medicago (secondary y-axis) based upon estimated synonymous distance measures. The curves are normal distribution densities (y-axis) for the statistically significant number of distributions. Beside each curve are the mean of synonymous distances under that curve as well as the correspondingly calculated coalesence estimates in million years ago (MYA).

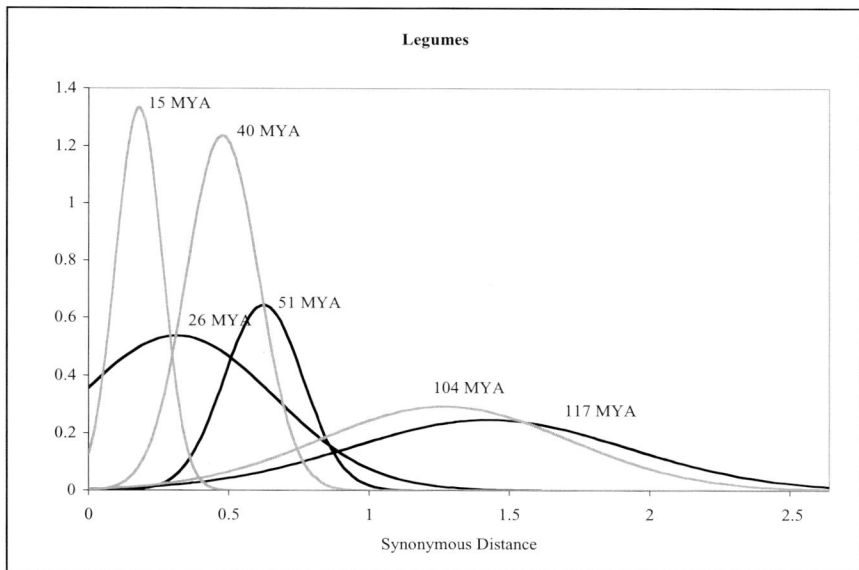

**FIGURE 2.** Normal distribution densities for both soybean and Medicago. Light grey curves correspond to soybean distributions. Black curves correspond to Medicago distributions. The dates next to each curve are the coalesence estimates in million years ago (MYA).

genome duplication is relatively recent. Our results suggest that after the divergence of soybean and Medicago, soybean underwent two major genome events (15 MYA and 40 MYA). Study of relationships within gene families has long suggested that soybean is an ancient polyploid (Lee and Verma, 1984; Hightower and Meagher, 1985; Grandbastien et al., 1986; Nielsen et al., 1989; Granger et al., 2002). Hybridization-based genetic maps are consistent with these findings (Shoemaker et al., 1996; Lee et al., 1999; Lee et al., 2001). Shoemaker et al. (1996) observed nested duplications, suggesting that one of the soybean genomes involved in the recent tetraploidization event could have undergone duplication prior to tetraploidization. This can be supported by the existence of the two peaks at 15 and 40 MYA, both occurring after the predicted divergence of soybean and Medicago.

Medicago probably underwent a major genome duplication event just after separation from soybean, at approximately 51 MYA. However, the 'event' at 26 MYA is represented as a broad peak (Figure 1) and may not represent a major event but may result from numerous chromosomal events spread over time. This would be consistent with the hypothesis that Medicago has not undergone a recent whole-genome duplication event. However, considering the 51 MYA event, Medicago is most likely an ancient polyploidy.

We should consider that a molecular clock is only a relative measure of time. An alternative hypothesis to the one proposed above is that the genomes of soybean and Medicago have very different evolutionary rates. The two most ancient peaks observed for each species might actually be events that occurred in concert with one another. Under this hypothesis, the 117 MYA and 51 MYA events in Medicago could coincide with the 104 MYA and 40 MYA events in soybean, respectively. These duplication events would have occurred prior to divergence of soybean and Medicago. Then soybean, approximately 15 MYA, underwent an independent large-scale event, whereas Medicago experienced ongoing small-scale duplications. Regardless, the two hypotheses both support the paleopolyploid history of soybean, with at least one early duplication event occurring in both soybean and Medicago prior to their divergence.

The maize genome is proposed to be a segmental allotetraploid (Gaut and Doebley, 1997). Based upon examination of 14 sequences, they estimated that the maize genome underwent a tetraploidization event approximately 11.4–16.5 MYA, with the beginning divergence of the two diploid progenitors occurring approximately 20.5 MYA (Gaut and Doebley, 1997).

Our analysis of 95 pairs of maize gene duplicates yields three normal distributions representing at least two large genome duplication events at approximately 16, 62, and 127 MYA (Figure 3). Although our results do not support the hypothesis of Gaut and Doebley (1997), a detailed examination of the histogram yielding the 16 MYA distribution results in two peaks with coalescence times corresponding to the earlier study (data not shown). However, these two peaks are not supported statistically. It is possible that the bimodal distribution observed by Gaut and Doebley was actually a single distribution, a single genome event.

The bimodal distributions of genetic distance between duplicate rice genes (Figure 3) are consistent with the established view of rice genome evolution. Rice is likely an ancient polyploid or the product of large-scale segmental duplications, e.g., between chromosomes 11 and 12 (Nagamura et al., 1995; Livingstone and Rieseberg, 2002; Yu et al., 2002). The sharp peak yielding a date of 6 MYA may be indicative of segmental duplications observed between chromosomes 11 and 12. The broad peak corresponding to a date of 59 MYA most likely represents a series of small-scale duplications occurring over time. Although clusters of gene pairs are observed under the broad distribution the treatment of each cluster as individual events is not statistically supported.

The cereals probably shared a common ancestor up until 55—70 MYA, after which the grasses began to evolve along separate pathways (Kellogg, 2001). At some time prior to 29 MYA, the rice lineage separated from the maize lineage (Zhang et al., 2001; Thomasson, 1987). The peak at 62 MYA in corn is very sharp and strongly supported statistically. Even though the peak representing the 59 MYA distribution in rice is supported as the most probable, it is shallow and broad and likely represents a series of small-scale events.

Two scenarios are possible from an analysis of this data: 1) rice and maize diverged prior to the 62 MYA event observed in maize, and rice underwent a series of small-scale duplications, or less likely, 2) rice and maize shared a common genome at approximately 59– 62 MYA, at which time that genome underwent a large-scale

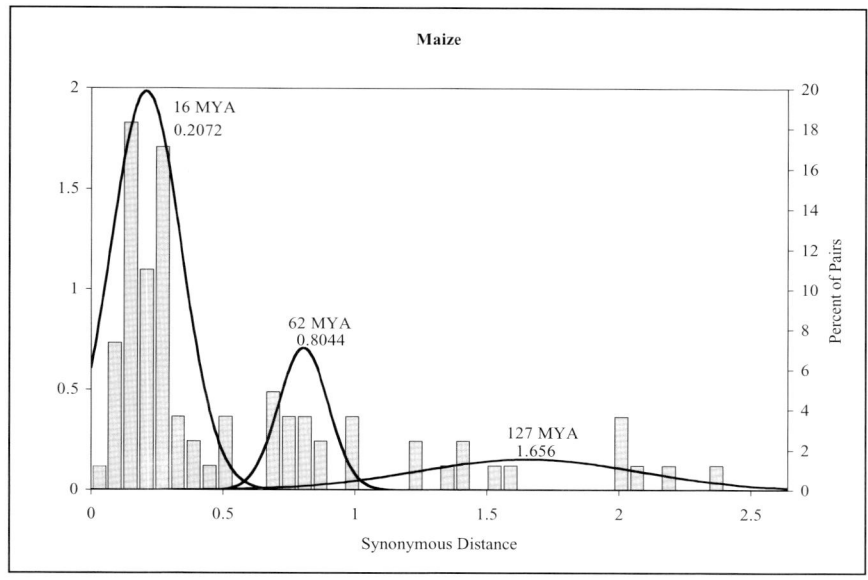

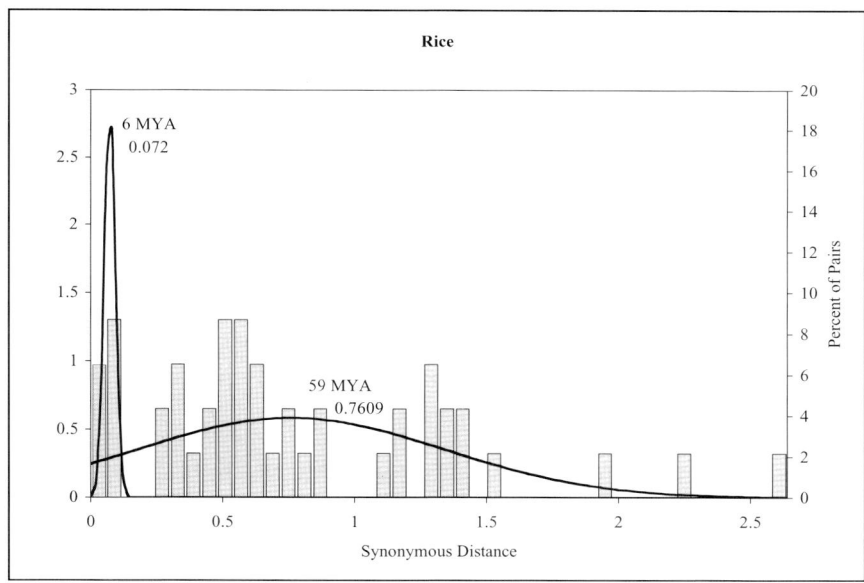

**FIGURE 3.** Histogram of the percent of identified pairs for both maize and rice (secondary y-axis) based upon estimated synonymous distance measures. The curves are normal distribution densities (y-axis) for the statistically significant number of distributions. Beside each curve are the mean of synonymous distances under that curve as well as the correspondingly calculated coalesence estimates in million years ago (MYA).

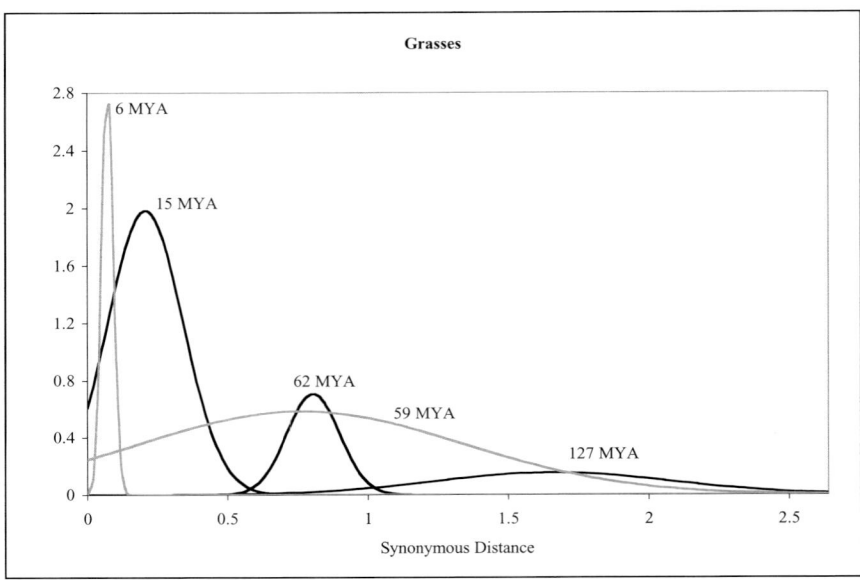

**FIGURE 4.** Normal distribution densities for both maize and rice. Light grey curves correspond to rice distributions. Black curves correspond to maize distributions. The dates next to each curve are the coalesence estimates in million years ago (MYA).

duplication event (Figure 4). If the second hypothesis is true, it would mean that the two genomes are undergoing radically different evolutionary trajectories. A larger sample size could resolve this issue.

A few caveats of distance measures should be addressed. First, an overestimation of "time since divergence" can occur due to asymmetrically bounded variables (non-elastic boundary at the present and an elastic boundary at the past) (Rodriguez-Trelles et al., 2002). Secondly, reduction in the precision of estimates of divergence can also be caused by variation in rates of sequence substitution (Wray, 2001) and rates of substitution may vary with functional differentiation, something often associated with gene duplication (Ohta 1994; Pickett and Meeks-Wagner 1995). Third, low numbers of ESTs in contigs can result in incomplete coverage of the coding region and can result in biased distance measures. Finally, ancient duplications become saturated at synonymous sites. This leads to potentially skewed estimates of the relative numbers of synonymous and nonsynonymous substitutions among duplicated genes in ancient duplications and may contribute to inaccuracies in the estimation of divergence times (Ohta, 1994). Several of these experimental errors can be minimized. Utilizing EST contigs with large numbers of EST's increases the probability of obtaining the full coding region. Also, as the number of gene sequences increases, as is possible with analysis of large datasets such as ours, the accuracy of the divergence estimates also increases (Nei et al., 2001).

## 6. POSITIVELY SELECTED GENES

Figures 5 and 6 depict a scatter plot of synonymous distances vs. nonsynonymous distances for all four species and allows us to determine the selective pressures the duplicated pairs experience. When the synonymous distance is greater than the nonsynonymous distance, the genes experience negative or purifying selection. Synonymous distances that are approximately equal to nonsynonymous distances are indicative of neutrality. Positive selection is indicated when the nonsynonymous distance is greater than the synonymous distance, i.e., more mutations result in protein sequence changes conservation of protein sequence. As seen in Figure 5, the vast majority of genes experience purifying selection. A closer examination of synonymous and nonsynonymous distance does identify genes showing positive selection (Figure 6).

Each gene of each pair was searched against the NCBI non-redundant database using the tblastx algorithm and an E cutoff value of $1 \times 10^{-10}$. The top five queries identified were then parsed from the blast output and classified into the appropriate first level MIPS (Munich Information Center for Protein Sequences) categories (Mewes et al., 2002). An additional category of nodulation/nitrogen fixation was added for legumes.

Table 3 shows the MIPS functional classification of each gene potentially under positive selection. There does not seem to be any correlation between the classifications across all species. However, the classes under positive selection are those most involved in cellular trafficking, communication, and control of

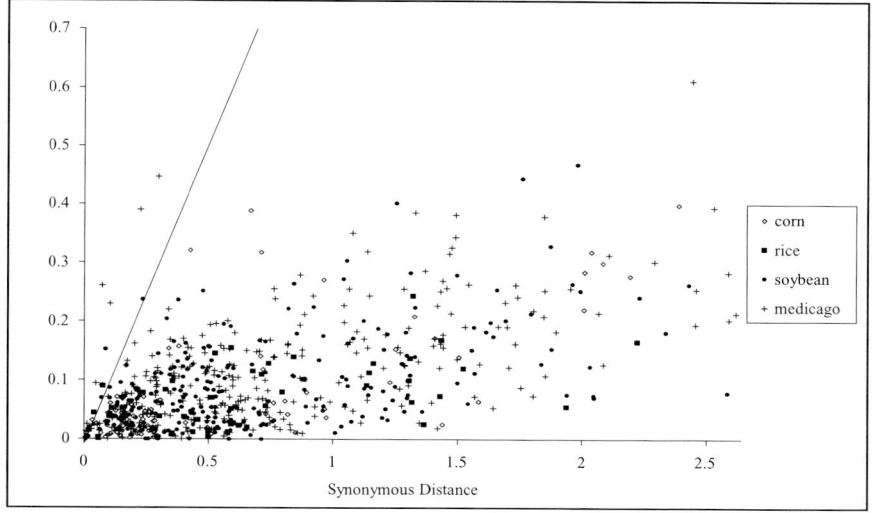

**FIGURE 5.** Nonsynonymous distance measures as a function of synonymous distance measures of all genes for all studied species. Each point is a duplicated gene pair. The line denotes neutral evolution, when synonymous distance is approximately equal to nonsynonymous distance.

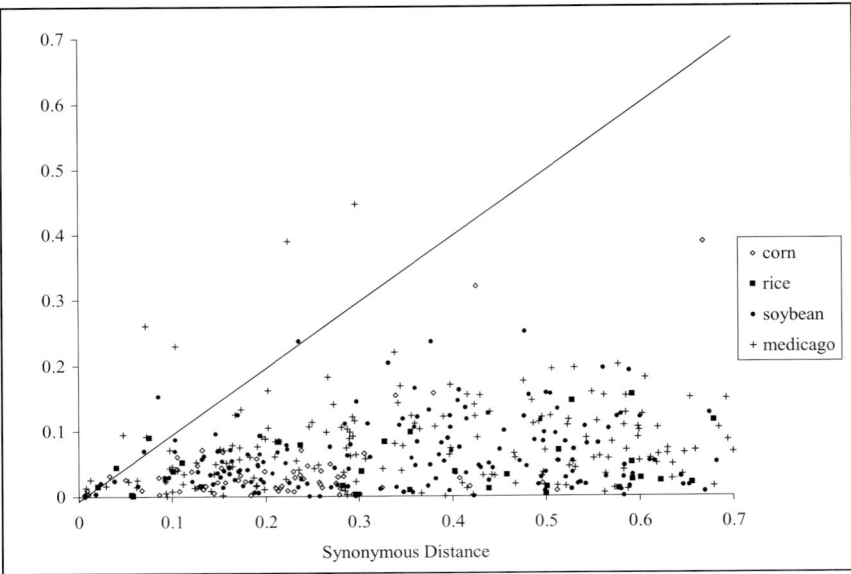

**FIGURE 6.** Nonsynoymous distance measures as a function of synonymous distance measures for more recently duplicated genes in all studies species. Each point is a duplicated gene pair. The line denotes neutral evolution.

gene expression. What is striking is that there are no structural genes found under positive selection in any of the species. This may be indicative of the cellular architecture being highly conserved while the underlying networks change more over time. Interestingly, in Medicago the largest number of positively selected

**Table 3**
**Functional Classification of Genes Under Positive Selection.**

| MIPS Classification | G. max | M. truncatula | Z. mays | O. sativa |
|---|---|---|---|---|
| Transcription | — | 1 | — | — |
| Protein Synthesis | — | — | — | — |
| Protein Destination | — | 1 | 1 | — |
| Transport Facilitation | 1 | — | — | — |
| Energy | — | — | — | — |
| Metabolism | — | — | — | — |
| Cell Communication/Signaling | — | — | — | 1 |
| Cell Growth/ Division/DNA Synthesis | — | — | — | — |
| Nodulation/Nitrogen Fixation | — | 3 | — | — |
| Unknown | — | 4 | — | — |
| Average Synonymous Distance[a] | 8.31 | 9.29 | 0.90 | 0.00 |
| Average Nonsynonymous Distance[b] | 14.6 | 17.27 | 0.98 | 2.12 |
| Ratio (Nonsynonymous/Synonymous)[c] | 1.76 | 1.86 | 1.09 | NA |

[a] Average of the synonymous distances of all genes under positive selection
[b] Average of the nonsynonymous distances of all genes under positive selection
[c] Ratio based upon the average synonymous and nonsynonymous distances

genes are involved in nodulation and nitrogen fixation. This may be the result of complementary changes in nodulation/nitrogen fixation genes in response to changes in the symbiotic bradyrhizobium. It should be noted that the Medicago EST libraries are largely biased to roots compared to the soybean libraries, whereas relatively few EST's were derived from soybean roots or nodules.

Considering the ratio of nonsynonymous to synonymous distances, the legumes have a larger ratio than the grasses. Possibly, the legume genes showing positive selection have experienced a greater divergence, or the positively selected genes in the grasses have diverged to a point of no longer being recognized as pairs. This may also be due to differences in rates of evolution, or the evolutionary trajectories between the grasses and the legumes.

## 7. SUMMARY

Discovering the evolutionary histories of complex genomes is not a trivial task. However, as more sequence information accumulates, the picture begins to become clearer. The genomes of both legumes and grasses show evidence of gene and genome duplication. The purpose of this study was to utilize publicly available EST data to determine the evolutionary genetic history of several major plant species. Duplicated transcripts were identified and analyzed to gain insight into the modes and mechanisms of gene and genome evolution processes and to estimate coalescence times for the duplicated genes.

Whole genome duplications are also subject to varying levels of selective pressure. Ahn and Tanksley (1993) suggested that after whole genome duplication chromosomal rearrangements accumulated in maize. In other words, genome duplication is followed by genome rearrangement. This was supported in a study involving recently synthesized polyploids in the Brassica (Song et al., 1995). Our results support these hypotheses by showing that most genes are not under positive selection after duplication but do indeed experience purifying selection. We do, however, observe some genes under positive selection, these being primarily involved in cellular signaling, transport, and transcription, but not structural functions.

After duplication, genes undergo one of several fates. Both may retain function, they may diverge to have different functions, or one may be lost while the remaining gene continues its original function (Pickett and Meeks-Wagner, 1995). Maintenance of duplicated sequences within paleopolyploids suggests the involvement of negative selective pressures in their evolution. We observed this in our study.

As suggested by Lynch and Conery (2000), genes experience a brief relaxation in purifying selective pressure immediately after duplication. During this period most copies are silenced with the few remaining paralogs experiencing strong purifying, or negative, selection. Further, it has been suggested that during this relaxation in selection pressure, duplicated genes affect the fitness of the organism, leading to a temporary advantage for gene duplication (Kondrashov et al., 2002). It has been hypothesized that the relaxation in purifying selection is the main driving force of duplicate gene evolution, while positive selection plays a less prominent role (Kondrashov et al., 2002). Kondrashov et al. also showed that among genes with similar levels of divergence, duplicated pairs evolved faster than single copy genes (2002).

## 8. REFERENCES

Ahn, S., and Tanksley, S.D., 1993, Comparative linkage maps of the rice and maize genomes, *Proc Natl Acad Sci U S A* **90**:7980–7984.

Altschul, S.F., Gish, W., Miller, W., Myers, E.W., and Lipman, D.J., 1990, Basic local alignment search tool, *J. Mol. Biol.* **215**:403–410.

Arumganthan, K., and Earle, E.D., 1991, Nuclear DNA content of some important plant species, *Plant Molecular Biology Reporter* **9**:208–218.

Blanc, G., Barakat, A., Guyot, R., Cooke, R., and Delseny, M., 2000, Extensive duplication and reshuffling in the *Arabidopsis* genome, *Plant Cell* **12**:1093–1101.

Blanc, G., Hokamp, K., and Wolfe, K.H., 2003, A recent polyploidy superimposed on older large-scale duplications in the *Arabidopsis* genome, *Genome Research* **13**:137–144.

Devos K.M., and Gale, M.D., 1997, Comparative genetics in the grasses, *Plant Molecular Biology* **35**: 3–15.

Doyle, J.J., and Luckow, M.A., 2003, The rest of the iceberg. Legume diversity and evolution in a phylogenetic context, *Plant Physiology* **131**:900–910.

Ewing R.M., Kahal A.B., Poirot O., Lopez F., Audic S., and Claverie J-M., 1999, Large-scale statistical analysis of rice ESTs reveal correlated patterns of gene expression, *Genome Research* **9**:950–959.

Frendo, P., Hernández M.J., Mathieu C., Duret, L., Gallesi, D., Van de Sype, G., Hárouart, D., and Puppo, A., 2001, A *Medicago truncatula* homoglutathione synthetase is derived from glutathione synthetase by gene duplication, *Plant Physiology* **126**:1706–1715.

Gaut, B.S., Morton, B.R., McCaig, B.M., and Clegg, M.T., 1996, Substitution rate comparisons between grasses and palms: synonymous rate differences at the nuclear gene Adh parallel rate differences at the plastid gene rbcL, *Proc Natl Acad Sci U S A S* **93**:10274–10279.

Gaut B.S., and Doebly J.F., 1997, DNA sequence evidence for the segmental allotetraploid origin of maize, *Proc Natl Acad Sci U S A S* **94**:6809–6814.

Gaut, B.S., Le Thierry d'Ennequin, M., Peek, A.S., and Sawkins, M.C., 2000, Maize as a model for the evolution of plant nuclear genomes, *Proc Natl Acad Sci U S A* **97**:7008–7015.

Goff, S.A., Ricke, D., Lan, T.-H., Presting, G., Wang, R., Dunn, M., Glazebrook, J., Sessions, A., Oeller, P., Varma, H., Lange, B.M., Moughamer T., Xia, Y., Budworth, P., Zhong, J., Miguel, T., Paszkowski, U., Zhang, S., Colbert, B., et al., 2002, A draft sequence of the rice genome (*Oryza sativa* L. ssp. *japonica*). *Science* **296**:92–100.

Grandbastien, M.A., Berry-Lowe, S., Shirley, B.W., and Meagher, R., 1986, Two soybean ribulose-1,5-bisphosphate carboxylase small subunit genes share extensive homology even in distant flanking sequences, *Plant Mol. Biol.* **7**:451–465.

Granger, C., Coryell, V., Khanna, A., Kein, P., Vodkin, L., and Shoemaker, R.C., 2002, Identification, structure, and differential expression of members of a BURP domain containing protein family in soybean, *Genome* **45**: 693–701.

Grant, D., Cregan, P., and Shoemaker R.C., 2000, Genome organization in dicots: genome duplication in *Arabidopsis* and synteny between soybean and *Arabidopsis*, *Proc Natl Acad Sci U S A* **97**:4168–73.

Gualtieri, G., and Bisseling, T., 2002, Microsynteny between the *Medicago truncatula* SYM2-orthologous genomic region and another region located on the same chromosome arm, *Theor Appl Genet* **105**:771–779.

Gyorgyey J., Vauber D., Jimenez-Zurdo J.I., Charon C., Troussard L., Kondorosi A., and Kondorosi E., 2000, Analysis of *Medicago truncatula* nodule expressed sequence tags, *Mol. Plant Microbe Interact.* **13**:62–71.

Helentjaris, T., Weber, D., and Wright, S., 1988, Identification of the genomic locations of duplicate nucleotide sequences in maize by analysis of restriction fragment length polymorphisms, *Genetics* **118**:353–363.

Hightower, R., and Meagher, R., 1985, Divergence and differential expression of soybean actin genes., *EMBO J.* **4**:1–8.

Huang, X., and Madan, A., 1999, CAP3: A DNA Sequence Assembly Program, *Genome Research* **9**: 868–877.

Kellogg, E.A., 2001, Evolutionary history of the grasses. *Plant Physiology* **125**:1198–1205.

Kimura, M., 1980, A simple method for estimating evolutionary rates of base substitutions through comparative studies of nucleotide sequences, *J. Mol. Evol.* **16**:111–120.

Kondrashov, F.A., Rogozin, I.B., Wolf, Y.I., and Koonin, E.V., 2002, Selection in the evolution of gene duplications, *Genome Biology* **3**:1–9.

Kowalski, S.P., Lan, T.H., Feldmann, K.A., and Paterson, A.H., 1994, Comparative mapping of *Arabidopsis thaliana* and *Brassica oleracea* chromosomes reveals islands of conserved organization, *Genetics* **138**:499–510.

Ku, H-M., Vision, T., Liu, J., and Tanksley, S.D., 2000, Comparing sequenced segment of the tomato and *Arabidopsis* genomes: Large-scale duplication followed by selective gene loss creates a network of synteny, *Proc Natl Acad Sci U S A* **97**:9121–9126.

Kulikova, O., Gualtieri, G., Geurts, R., Kim, D.J., Cook, D.R., Huguet T., De Jong, H., Franz, P.F., and Bisseling, T., 2001, Integration of the FISH pachytene and genetic maps of *Medicago truncatula*, *Plant J* **27**:49–58.

Lackey, J.A., 1980, Chromosome numbers in the Phaseoleae (Fabaceae:Faboideae) and their relation to taxonomy, *American Journal of Botany* **67**:595–602.

Lee, J.M., Bush A., Specht J.E., and Shoemaker R., 1999, Mapping duplicate genes in soybean, *Genome* **42**:829–836.

Lee, J.M., Grant, D., Vallejos C.E., and Shoemaker R., 2001, Genome organization in dicots. II. *Arabidopsis* as a 'bridging species' to resolve genome evolution events among legumes, *Theor. Appl. Genet.* **103**:765–773.

Lee, J.S., and Verma, D.P.S., 1983, Chromosomal arrangement of leghemoglobin genes in soybean, *Nucleic Acids Research* **11**:5541–5553.

Li, W.-H., Wu, C.-I., and Luo, C.-C., 1985, A new method for estimation synonymous and nonsynonymous rates for nucleotide substitution considering the relative likelihood of nucleotide and codon changes, *Mol. Biol. Evol.* **2**:150–174.

Li, W.H., 1993, Unbiased estimation of the rates of synonymous and nonsynonymous substitution, *J. Mol. Evol.* **36**:96–99.

Livingstone, K., and Rieseberg, L.H., 2002. Rice genomes: a grainy view of future evolutionary research. *Current Biology* **12**:470–471.

Lunde C.F., Morrow D.J., Roy L.M., and Walbot V., 2003. Progress in maize gene discovery: a project update, *Funct. Integr. Genomics* **3**:25–32.

Lynch, M., and Conery, J.S., 2000, The evolutionary fate and consequences of duplicate genes, *Science* **290**:1151–1155.

McCouch, S.R., Kochert, G., Yu, Z.H., Wang, Z.Y., Khush, G.S., Coffman, W.R., and Tanksley, S.D., 1988, Molecular mapping of rice chromosomes, *Theor Appl Genet* **76**:815–829.

McGrath, J.M., Jancso, M.M., and Pichersky, E., 1993, Duplicate sequences with a similarity to expressed genes in the genome of *Arabidopsis-thaliana Theor Appl Genet* **86**:880–888.

Mewes, H.W., Frishman, D., Guldener, U., Mannhaupt, G., Mayer, K., Mokrejs, M., Morgenstern, B., Munsterkotter, M., Rudd, S., and Weil, B., 2002, MIPS: a database for genomes and protein sequences, *Nucleic Acids Res* **30**:31–4.

Moore, G., Devos, K.M., Wang, Z., and Gale, M.D., 1995, Cereal genome evolution, *Current Biol* **5**:737–739.

Nagamura, Y., Inoue, T., Antonio, B.A., Shimano, T., Kajiya, H., Shomura, A., Lin, S.Y., Kuboki, Y., Harushima, Y., Kurata, N., Minobe, Y., Yano, M., and Sasaki, T., 1955, Conservation of duplicated segments between rice chromosome-11 and chromosome-12, *Breeding Science* **45**:373–376.

Nei, M., Xu, P., and Glazko, G., 2001, Estimation of divergence times from multiprotein sequences for a few mammalian species and several distantly related organisms, *Proc Natl Acad Sci USA* **98**:2497–2502.

Nielsen, N.C., Dickinson, C., Cho, T.J., Thanh, V.H., Scallon, B.J., et al., 1989, Characterization of the glycinin gene family in soybean, *Plant Cell* **1**:313–328.

Ohno, S., 1970, *Evolution by Gene Duplication*, Springer-Verlag, New York.

Ohta, T., 1994, Further examples of evolution by gene duplication revealed through DNA sequence comparisons, *Genetics* **138**:1331–1337.

Pamilo, P., and Bianchi, N.O., 1993, Evolution of the Zfx and Zfy genes: rates and interdependence between the genes, *Mol. Biol. Evol.* **10**:271–281.

Pickett, F., and Meeks-Wagner, R., 1995, Seeing double: appreciating genetic redundancy, *Plant Cell* **7**:1347–1356.

Rhoades, M.M., 1951, Duplicate genes in maize, *American Naturalist* **85**:105–110.

Rice, P., Longden, I., and Bleasby, A., 2000, EMBOSS: The European Molecular Biology Open Software Suite. *Trends in Genetics* **16**:276–277.

Rodriguez-Trelles, F., Tarrio, R., and Ayala, F., 2002, A methodological bias toward overestimation of molecular evolutionary time scales, *Proc Natl Acad Sci U S A*. **9**:8112–8115.

Salse, J., Biegu, B., Cooke, R., and Delseny, M., 2002, Synteny between *Arabidopsis thaliana* and rice at the genome level: a tool to identify conservation in the ongoing rice genome sequencing project, *Nucleic Acids Research* **30**:2316–2328.

Shoemaker, R., Polzin, K., Labate, J., Specht, J., Brummer, E.C., Olson, T., Young, N., Concibido, V., Wilcox, J., Tamulonis, J., Kochert, G., and Boerma, H.R., 1996, Genome duplication in soybean (*Glycine* subgenus *soja*), *Genetics* **144**:329–338.

Shoemaker, R.C., Keim, P., Vodkin, L., Retzel, E., Clifton, S.W., Waterston, R., Smoller, D., Coryell, V., Khanna, A., Erpelding, J., Gai, X.W., Brendel, V., Raph-Schmidt, C., Shoop, E.G., Vielweber, C.J., Schmatz, M., Pape, D., Bowers, Y., Theising, B., Martin, J., Dante, M., Wylie, T., and Granger, C., 2002, A compilation of soybean ESTs: generation and analysis, *Genome* **45**:329–338.

Simillion, C., Vandepoele, K., Van Montagu, M.C.E., Zabeau, M., and Van de Peer, Y., 2002, The hidden duplication past of *Arabidopsis thaliana*, *Proc Natl Acad Sci U S A* **99**:13627–13632.

Song, K., Lu, P., Tang, K., and Osborn, T.C., 1995, Rapid genome change in synthetic polyploids of *Brassica* and its implications for polyploidy evolution, *Proc Natl Acad Sci U S A* **92**:7719–7723.

Stebbins, G.L., 1950, *Variation and Evolution in Plants*, Columbia University Press, New York.

Stebbins, G.L., 1966, Chromosomal variation and evolution, *Science* **152**:1463–1469.

Thomasson, J.R., 1987, Fossil grasses: 1820–1987. In: Grass *systematics and evolution*, Soderstrom, T.R., Hilu, K.H., Campbell, C.S., Barkworth, M.E., eds., Smithsonian Institution Press, pp 159–167.

Vision, T.J., Brown, D.G., and Tanksley, S.D., 2000, The origins of genomic duplications in *Arabidopsis*, *Science* **290**:2114–2117.

Wolfe, K.H., Gouy, M., Yang, Y.W., Sharp, P.M., and Li, W.H., 1989, Date of the monocot-dicot divergence estimated from chloroplast DNA sequence data, *Proc Natl Acad Sci U S A* **86**:6201–5.

Wray, G.A., 2001, Dating branches on the tree of life using DNA, *Genome Biol.* **3**:1–7.

Wisconsin Package Version 10, Genetics Computer Group (GCG), Madison, Wisc.

Xing, L., and Brendel, V., 2000, MuSeqBox: a program for multi-query sequence BLAST output examination, *Bioinformatics* **17**, 744–745

Yan, H.H., Mudge, J., Kim, D.-J., Shoemaker, R.C., Cook, D.R., and Young, N.D., 2003, Estimates of conserved microsynteny among the genomes of *Glycine max, Medicago truncatula, and Arabidopsis thaliana. Theor. Appl. Genet.* (in press).

Yang, Y.-W., Lai, K.-N., Tai, P.-Y., and Li, W.-H., 1999, Rates of nucleotide substitution in angiosperm mitochondrial DNA sequences and dates of divergence between *Brassica* and other angiosperm lineages, *Journal of Molecular Evolution* **48**:597–604.

Yu, J., Hu, S., Wang, J., Wong, G.K.-S., Li, S., Liu, B., Deng, Y., Dai., L., Zhou, Y., Zhang, X., Cao, M., Liu, J., Sun, J., Tang, J., Chen, Y., Huang, X., Lin, W., Ye, C., Tong, W., cong, L., Huang, X., Li, W., Li, J., et al., 2002, A draft sequence of the rice genome (*Oryza sative* L. ssp. *indica*), *Science* **296**:79–92.

Zhang, L., Kosakovsky, S., and Gaut, B.S., 2001, A survey of the molecular evolutionary dynamics of twenty-five multigene families from four grass taxa, *Journal of Molecular Evolution* **52**:144–156.

*Chapter 13*

# A Biologist's View of Systems Integration Systems Biology

## The Pathogen Portal Project

R. Lathigra, Y. He, R.R. Vines, E.K. Nordberg,
and B.W.S. Sobral

## 1. INTRODUCTION

Biological data sets are growing exponentially as a result of whole genome sequencing projects and other high-throughput functional data generated by transcriptomic, proteomic and metabolomic technologies (Eckart and Sobral, 2004). Many of the pathogen genomes belonging to the Centers for Disease Control (CDC) and the National Institute of Allergy and Infectious Diseases (NIAID) priority lists have been sequenced and annotated, however, their sequences and annotations require continuous updating as biological knowledge increases. To date, there are no agreed upon standards in the bioinformatics community for data display and interoperation and, as a result, data are difficult to gather, underutilized, and suboptimally organized. Data have to be formatted and reformatted for analysis and management during the process of annotation while data analysis requires multiple tools that are not interoperable. Meanwhile, life scientists need to process even more data and convert them into information about genetics and biochemistry. Not all individual scientists have the necessary information technology (IT)

**R. Lathigra, Y. He, R.R. Vines, E.K. Nordberg, and B.W.S. Sobral**    Virginia Bioinformatics Institute, Virginia Polytechnic Institute and State University, Blacksburg, Virginia 24061-0477

*Genome Exploitation: Data Mining the Genome*, edited by J. Perry Gustafson, Randy Shoemaker, and John W. Snape.
Springer Science + Business Media, New York, 2005.

background and support to fully exploit the available information. In order to maximize the knowledge gained through new technologies, a comprehensive national strategy should be established to integrate pathogen-related data from different sources and to provide mechanisms for efficient data analysis, visualization, and sharing (Eckart and Sobral, 2004).

CDC lists more than 40 bacterial or viral pathogens according to their priority for bioterrorism potential (http://www.bt.cdc.gov/agent/agentlist-category.asp). NIAID provides a similar list for research priority including emerging infectious diseases (http://www.niaid.nih.gov/biodefense/bandc_priority.htm). Although the CDC and many other websites contain general background information about these pathogens and diseases, the information is in flat-file format, not machine readable, and sometimes hard to query for specific information. Some information, such as molecular mechanisms for pathogenesis, is not always available at these sites.

Many excellent pathogen-related databases and software tools are currently available, however, these databases and software tools are not typically interoperable across pathosystems or data types and furthermore many good software tools use a Unix command line interface that tends to inhibit many biologists from using the programs. Hence, a framework that allows interoperation of pathogen data and tools from different sources would be of value in infectious diseases research and in the development of associated counter-measures.

The Pathogen Portal Project (PathPort) at Virginia Bioinformatics Institute (VBI) is a collaborative data collection and software development effort to acquire, curate, and provide up-to-date pathogen information to infectious disease researchers and to provide a method for interoperation of data and tools from different sources. The PathPort project goal is to provide a bioinformatics infrastructure for data acquisition, storage, analysis, visualization, and dissemination of host-pathogen data, thereby helping infectious diseases research advances. The PathPort infrastructure is open, flexible, and easily extendable to support evolving community data and tools. This was achieved by developing VBI's ToolBus client-side interconnect technology (Eckart and Sobral, 2004; Eckart, et al., 2002) and by leveraging Web Services (Gardner, 2001) based on eXtensible Markup Language (XML) (Achard, et al., 2001). The first data and tools available through PathPort are genomic and functional genomic data.

PathPort runs individual Web Services (Gardner, 2001) on the server side and ToolBus as a client-side interface to interconnect different data and tools. ToolBus is platform independent due to its Java2 implementation, providing simultaneous support for different operating systems including Linux/Solaris, Windows, and Mac OS X. Data communication between different tools and services is based on community XML standards, a universal format for structured documents and data on the Web (Gardner, 2001) that is actively supported by the Interoperable Informatics Infrastructure Consortium (I3C) (Interoperable Informatics Infrastructure Consortium; http://www.i3c.org) among others. ToolBus supports the dynamic discovery of web-services and loading plug-ins, as well as the saving and loading of work sessions, which can be shared with colleagues via e-mail as .zip file attachments. PathPort provides an open framework to support XML-based data and tool interoperability and its application to molecular and cellular data sets for key host-pathogen-environment interactions. The Path-Port/ToolBus software and source code are freely available for non-commercial uses at http://www.vbi.vt.edu/pathport.

## 2. ILLUSTRATION OF PATHPORT

Two illustrations of the PathPort system are shown below. The first demonstrates how the tools of PathPort system were used to further our own research on *Brucella,* a CDC Category B pathogen that causes brucellosis in humans and other mammals. The second pertains to pathogen background information.

### 2.1. Use of PathPort for Genome Annotation and Comparison

Genome annotation can begin once the sequence of a genome has been determined. Genome annotation is a process of identifying genes and assigning features, functions, and attributes to specific DNA sequences *in silico*, using various database searches and prediction programs. Following analysis by gene prediction programs, BLAST/FASTA similarity alignments are run to identify orthologs and paralogs. Identified hits are subjected to multiple sequence alignment, followed by phylogenetic analysis to determine relatedness and ancestry. Deduced protein sequences are also analyzed against the Cluster of Orthologous Groups (COG) (Tatusov, et al., 2000) database, PFAM (Bateman, et al., 2002) to identify conserved protein domains, and Gene Ontology (GO) (Consortium GO, 2001) categories to determine the putative function of a gene product. Differences in the genomes of related organisms can be elucidated by comparative genomics approaches. Identified differences can be of great value in the evaluation of proteins as candidates for novel vaccines and for generation of new molecular diagnostic techniques.

This approach to annotation was used to discover genes uniquely present on chromosome II of *Brucella suis* by using several, interoperable PathPort analysis tools. *Brucella* spp. are intracellular facultative pathogens that cause brucellosis in humans and many mammals. The *B. melitensis* and *B. suis* genomes have been sequenced while the *B. abortus* genome has been sequenced and its annotation is in progress.

As outlined above, we began genome annotation by analysis with the gene prediction program GLIMMER (Delcher, et al., 1999), followed by BLAST/FASTA analysis, multiple sequence alignments, and phylogenetic tree analysis. PathPort provides a number of gene prediction programs as web-services to perform genome annotation. "Features" of the sequence were identified during annotation and these were stored in a VBI-designed, Oracle relational database. These features then become available through the VBI's Distributed Annotation System (Eckart and Sobral, 2004) (Dowell, et al., 2001) (VBI-DAS) viewer (Figure 1). Different feature types appear in different colors based on choices made by the user. The positions

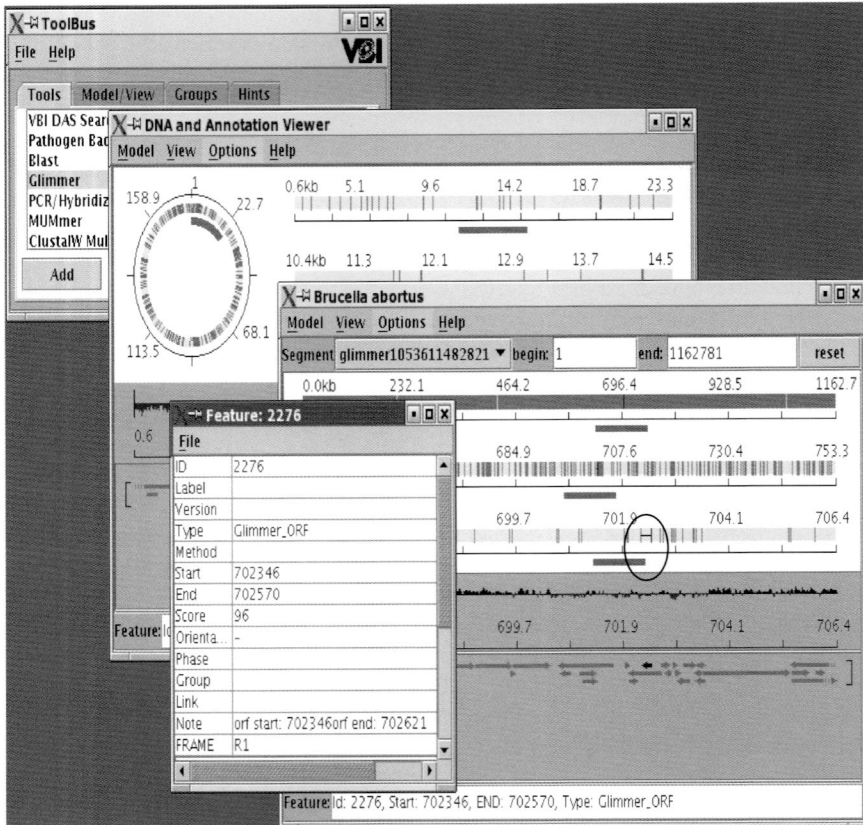

**FIGURE 1.** Genome annotation. Glimmer[9] was selected as shown in the background screenshot as a genome annotation tool. Glimmer gene predictions were then displayed in a DAS-viewer. The DNA and Annotation viewer shows the genome represented as a circle on the left-hand side. An internal arc can be dragged around the circle. The region covered by the arc is shown on the right of the circle as a linear segment. In this example with *Brucella abortus*, the position of an ORF as predicted by Glimmer is noted as a horizontal bar (circled near position 701.9). Clicking on the location of the ORF spawns a window showing details about the features associated with the ORF. Features can be added, edited, or deleted as part of the genome annotation process.

and directions of transcription-translation are depicted as solid colored arrows pertaining to the feature. Any linear and circular sequences, with XML formats, that match DAS.dtd can be viewed using the VBI-DAS viewer. On circular genomes, an internal arc within a circle can be dragged around the circle for selecting specific regions to display as linear segments in the same viewer. A graphical interface showing regions of identity/similarity helps annotators assign functions based on computer predicted functions. As mentioned previously, clicking on the ORF will spawn a window where all the features for the ORF can be added as part of the annotation process.

Individual features include a promoter location, ribosome-binding site, start and stop position of a gene, or an open reading frame (ORF). Clicking on the feature tag spawns a window that displays more details. The architecture of ToolBus is modular so several tools can be used simultaneously. Thus, genes predicted by gene prediction programs and viewed in the DAS viewer can be "grabbed, dragged, and dropped" into fields for BLAST/FASTA analysis without reformatting the data. BLAST/FASTA results can be selected, dragged and dropped into fields for "multiple sequence alignment" (MSA) either at the DNA level or the protein level. A hydrophobicity plot of each protein can be viewed from the MSA.

Differences between chromosomes II of *B. abortus* and *B. suis* were discovered using comparative genome analysis tools. These two chromosomes were compared using MUMmer (Delcher, et al., 2002), and results were displayed in a Parallel Sequence Comparison viewer or a Perpendicular Sequence Comparison viewer (a MUMmer view) (Figure 2). Areas where gene inversions or deletions may have occurred were easily discernible on the Perpendicular viewer. Selecting a position of gene inversion in one viewer is automatically displayed in the alternate comparison viewer through a built-in dynamic linking of the two comparison viewers. Clicking on a region of the comparison viewer spawns a window showing details of relatedness. As shown in Figure 2, presence of unique genes can be easily identified by the lack of homology lines linking the two genomes in the Parallel viewer. Similarly, lack of homology due to presence or absence of a gene can be identified by a break in the diagonal line seen in the Perpendicular viewer (Figure 2). Using such whole genome comparison tools in conjunction with BLAST/FASTA analysis, we identified 28 genes that are uniquely present on *B. suis* chromosome II as compared to *B. abortus*. Molecular diagnostic technologies rely on the ability to identify unique genes present in a pathogen and designing gene specific primers for PCR amplification. Dragging and dropping gene sequences into the Primer3 (Rozen and Skaletsky, 2000) design software available in PathPort can generate primers and probes for any gene. An example of primers designed to PCR amplify one of the unique genes of *B. suis* is shown in Figure 3. The presence or absence of a gene(s) in a virulent and avirulent organism (e.g., *E. coli* O157:H7 EDL933 and *E. coli* MG1655, respectively) can be investigated through whole genome comparison tools as shown above. Genomic sequence comparisons can be performed on a sequence of interest using the VBI-DAS database, which contains genome sequences for Archaea, prokaryotes, eukaryotes, and viruses that have been downloaded from the National Center for Biotechnology Information (NCBI genomes. http://www.ncbi.nlm.nih.gov/Genomes/index.html).

## 2.2. PathPort As a Pathogen Information Resource

In preparation for a bioterrorism attack, personnel involved in the ensuing response would benefit from preemptively knowing the properties of the organisms that can be used as Biological Warfare Agents (BWA), what procedures can be used for rapid identification of the BWA, and what prevention measures should be implemented to contain the outbreak of the associated disease. An educational

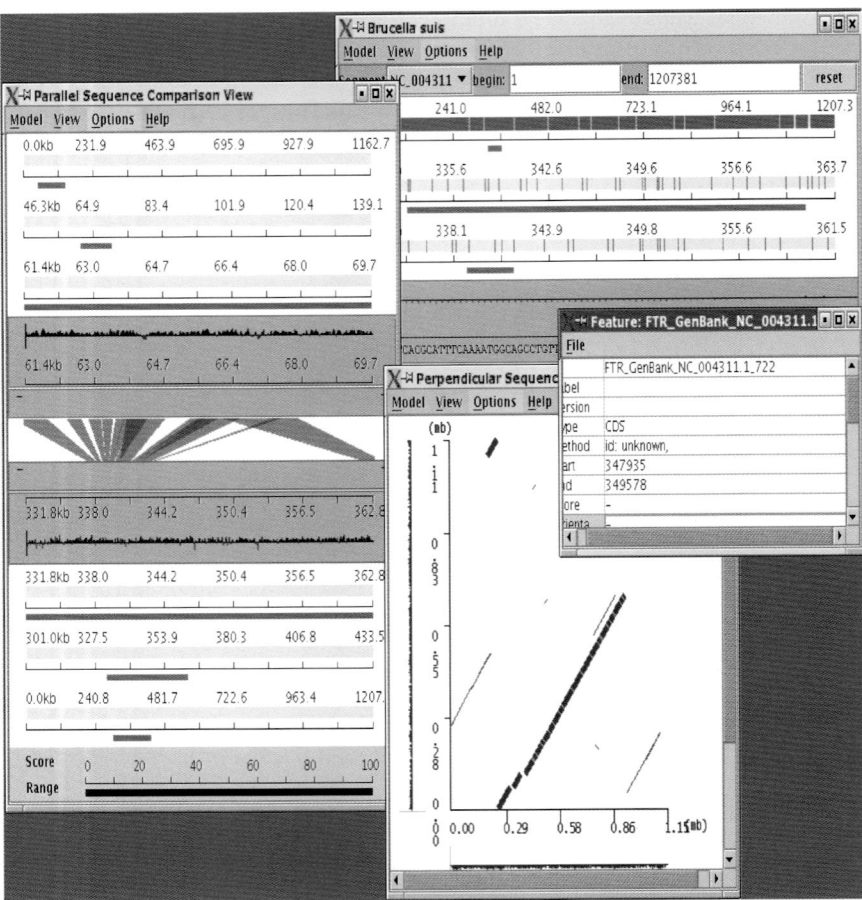

**FIGURE 2.** Genome Comparison. Chromosome II of *Brucella suis* and *Brucella abortus* are selected for comparison using either the Parallel Sequence Comparison View (left hand side) or the Perpendicular Sequence Comparison View (MUMmer view) (right hand side). Genome differences in the Perpendicular view are shown as breaks within the diagonal line, whereas they are shown as white space in the Parallel view. The exact location of a unique region on Chromosome II of *B. suis* is shown as a feature in the foreground screenshot.

resource containing pertinent information about pathogens with potential use as BWAs would be very useful for healthcare workers, emergency response teams and relevant Federal agencies.

PathPort contains a Pathogen Background Information (PathInfo) tool and its related PathInfo viewer (Figure 4A). PathInfo currently includes publicly available information for over 20 pathogens (Table 1). PathInfo contains sections on taxonomy, physical characteristics, epidemiology, host information, transmission mode, carriers, and laboratory procedures for rapid diagnosis of the pathogen. This information has been collected and annotated by a team of graduate students and

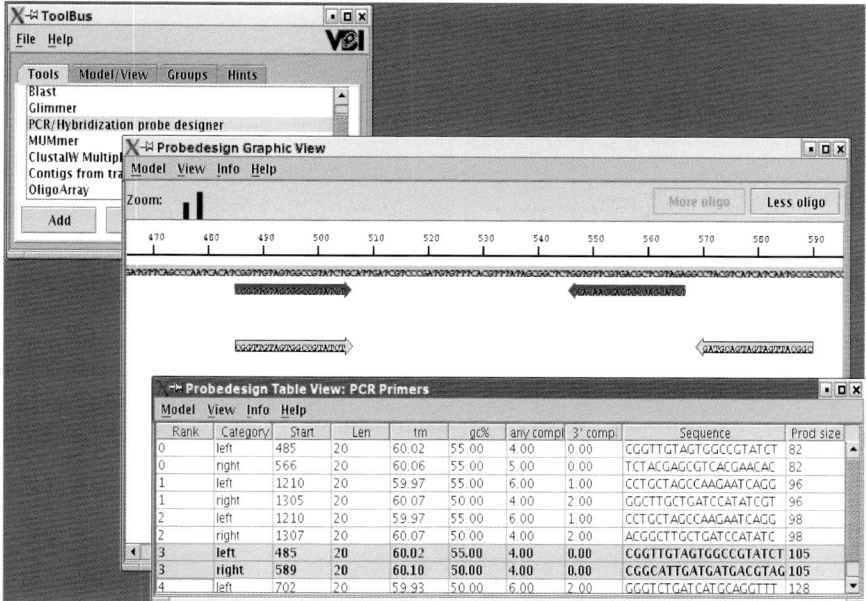

**FIGURE 3.** Probe design. The probe design model/view is based on Primer3[9]. Probe design results can be viewed in both graphic and table views. The graphic view provides an intuitive graphic representation of the probe design result. It shows the length and sequence of the target, the position, length, and sequence of the PCR primers and/or hybridization probes. The table view displays the same information in a tabular format and includes details about melting temperature and other parameters as selected by the user during primer design.

doctoral-level biologists who will also regularly update it to ensure that the latest data are presented. Furthermore, before being made available to the public, the quality and comprehensive nature of the information for each pathogen is verified by an "outside expert" in the field. A major advantage of PathInfo is its ability to integrate information from diverse sources, allowing for a complete compilation and computer access of pathogen information at one site. Resources being surveyed for information include: peer-reviewed journals, books, and websites. The information in PathInfo is thoroughly referenced, allowing the user to review the source of the data and for quick access, "hotlinks" to website references (or PubMed abstract in the case of journal articles) are used in PathInfo when possible. Where applicable, photographic images and diagrams are incorporated in PathInfo (with the requisite permission from publishers/owners) to more fully illustrate important points.

PathPort biologists also collect, annotate, and review information regarding pathogenesis and host/pathogen interactions for these pathogens at the molecular level. This information is available through the PathPort Protein Network (ProNet) viewer (Figure 4B). This curated information conforms to a defined XML format (http://www.vbi.vt.edu/pathport/xml/molecules/molecules.dtd) that can be downloaded and viewed by users. The network of proteins involved in

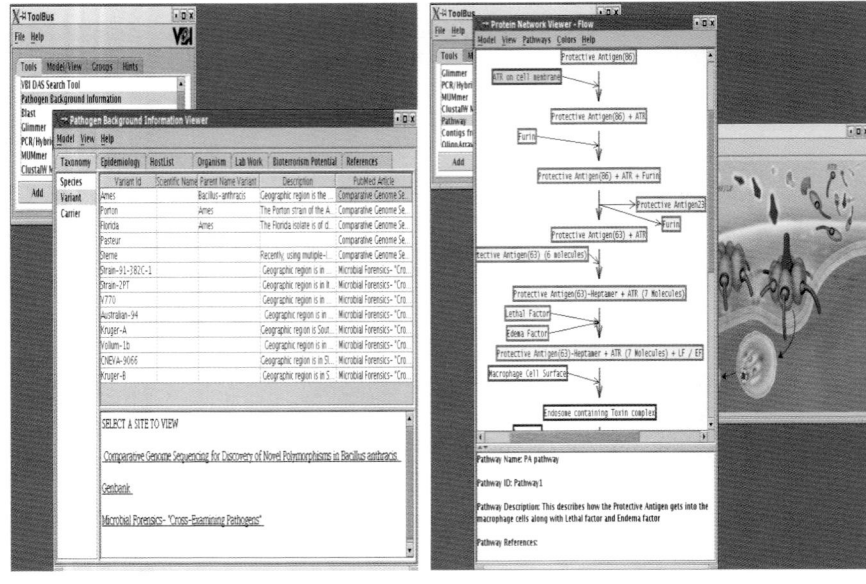

A                                                    B

**FIGURE 4.** PathInfo and ProNet. (4A). The "Pathogen Background Information" tool is highlighted in the background screenshot. The screenshot in the foreground shows several topics for which background information is available as tabs. The left panel provides the hierarchy structure of the selected tab information. The right panel shows the detailed information for a specific topic selected by clicking from the left panel. (4B). Protein-protein interactions are depicted in a pathway form. Pathway tool (selected) is shown in the background screenshot. The ProNet viewer (foreground screenshot) shows interactions between bio-objects (proteins), with location noted, in pathways as nodes and directed links between bio-objects. When available, pictorial representations of the pathway are provided in ProNet (e.g. right-most screenshot). The image shows the internalization of the lethal factor of *B. anthracis* into the cytoplasm of a host cell.

pathogenesis (both pathogen and host) is illustrated in a diagram as a stepwise progression. Clicking on the last entity of the protein network spawns another window to depict sub-interactions if they are present. Each entity of the network is color coded to indicate its cellular location in the pathway (for example, extracellular, intracellular, inside the Golgi apparatus, endoplasmic reticulum). Clicking on any entity in the pathway reveals additional information (its own pane) about the protein as well as related bibliography. Clicking on the reference identifier dynamically links the user to PubMed for easy retrieval of the reference.

From a single session of reviewing pathogen background information, the user is able to gather detailed information pertaining to the organism, the associated disease, diagnostic methods, and the mechanism of pathogenesis. This type of system enables researchers to readily identify proteins that could be targets for immune intervention or for small molecule inhibitors or for rapid diagnostics based on either immunological or nucleic acids based technologies.

## Table 1
## Pathogen Information Currently Available in PathPort and the Status of Curation (Eckart and Sobral, 2004).

| Pathogen | Associated Disease | CDC / NIAID Category | Curation Status In process | Final[2] |
|---|---|---|---|---|
| **Prokaryote** | | | | |
| *Bacillus anthracis* | Anthrax | CDC-A / NIAID -A | | ✓ |
| *Brucella* spp. (*B. melitensis*, *B. abortus*, and *B. suis*) | Brucellosis | CDC-B / NIAID-B | | ✓ |
| *Burkholderia mallei* | Glanders | CDC-B / NIAID-B | ✓ | |
| **Burkholderia pseudomallei** | Meliodosis | CDC-B / NIAID-B | ✓ | |
| *Chlamydia psittaci* | Psittacosis | CDC-B | ✓ | |
| *Clostridium botulinum* | Botulism | CDC-A / NIAID-A | ✓ | |
| *Coxiella burnetii* | Q fever | CDC-B / NIAID-B | ✓ | |
| *E. coli* O157:H7 | HUS, diarrhea | CDC-B / NIAID-B | | ✓ |
| *Francisella tularensis* | Tularemia | CDC-A / NIAID-A | | ✓ |
| *Mycobacterium tuberculosis* (MDR) | Tuberculosis | NIAID-C | ✓ | |
| *Rickettsia prowazekii* | Typhus fever | CDC-B / NIAID-B | ✓ | |
| *Salmonella* spp. | Diarrhea | CDC-B / NIAID-B | | ✓ |
| *Shigella dysenteriae* | Diarrhea | CDC-B / NIAID-B | | ✓ |
| *Vibrio cholerae* | Cholera | CDC-B / NIAID-B | ✓ | |
| *Yersinia pestis* | Plague | CDC-A / NIAID-A | | ✓ |
| **Eukaryote** | | | | |
| *Coccidiodes immitis* | Valley fever | – | ✓ | |
| *Cryptosporidium* spp. | Diarrhea | CDC-B / NIAID-B | ✓ | |
| *Phytophthora infestans* | Potato/tomato blight | – | ✓ | |
| *Phytophthora sojae* | Soybean root rot | – | ✓ | |
| *Plasmodium falciparum* | Malaria | – | ✓ | |
| **Virus** | | | | |
| *Crimean Congo Hemorrhagic Fever* | Hemorrhagic fever | NIAID-C | ✓ | |
| Eastern Equine Encephalitis | Encephalitis | CDC-B / NIAID-B | ✓ | |
| Hemorrhagic Fever Viruses[3] | Hemorrhagic fever | CDC-A / NIAID-A | | ✓ |
| Highly Pathogenic (HP) Avian Influenza virus | Influenza | – | ✓ | |
| Powassan | Encephalitis | – | ✓ | |
| Variola major / minor | Smallpox | CDC-A / NIAID-A | ✓ | |
| Venezuelan Equine Encephalitis | Encephalitis | CDC-B / NIAID-B | ✓ | |
| Yellow Fever | Yellow fever | NIAID-C | ✓ | |

## 3. OTHER PATHPORT TOOLS FOR FUNCTIONAL ANNOTATION

Identification of genes expressed by the pathogen and the host and the contribution of the environmental signals will vastly increase our knowledge of the dynamic interaction of the specific pathosystems. Genes identified to be specifically expressed in an *in vivo* environment could be targets for intervention strategies either as vaccine candidates or targets for small molecule inhibitors. Transcription analysis can increase our understanding of how a pathogen responds to both *in vitro* and *in vivo* specific stimuli (Conway and Schoolnik, 2003). In combination with proteomics and metabolomics data, a clearer picture of the inner workings of cells and pathogens is likely to emerge. In the context of gene regulatory networks, such network hypotheses can also be used to support functional annotations of genomes. Very few pathogen-specific gene chip arrays are commercially available, therefore, researchers are currently relying on pathogen-specific oligo-based microarrays. In future months, software including OligoArray (Rouillard, et al., 2002) will be provided within PathPort to facilitate microarray design. Corresponding tools for visualization and analysis of microarray data will also be incorporated. A relational database is being developed with PathPort to store all of the experimental microarray data enabling storage of large microarray data sets and comparison between different microarray groups. To confirm results obtained from transcriptional profiling experiments, RT-PCR can be used. To design PCR primers for this purpose, the Primer3[12]-based primer/hybridization oligo design tools in PathPort are currently fully implemented.

The connections between gene regulatory networks, the proteins being expressed at any given point in time, and the metabolites being generated during host-pathogen-environment interactions have to be established. It is becoming increasingly clear that such connections could lead to new discoveries and the establishment of new pathways for protein interactions. How can these types of data be mined to establish such connections? Special plug-in tools called group suggestors are available through PathPort/ToolBus allowing the user to search for connections between different data types. Interesting data associations are suggested as possible ToolBus groups and these groupscan be compared via user generated Venn diagrams that allow the elucidation of commonalities and differences between groups in negative and positive visual spaces. Such group suggestors are likely to play a significant role in the search for new interaction pathways and further the understanding of the interrelationship between seemingly disparate data.

## 4. DISCUSSION

The detailed PathPort/ToolBus framework has been described previously (Eckart and Sobral, 2004; Eckart, et al., 2002). Web-services and XML are utilized in PathPort to support remote data access and analysis, while client-side Java applications are performed for information visualization. Other approaches have also been developed based on centralization of one or more aspects of the system workflow. For instance, Grid middleware such as Avaki (Grimshaw, 2003) and Globus (Foster, and Kesselman, 1998) use a single unified virtual computer system to share data and resources among a variety of computer systems and organizations. Web-based portals such as BioASP (BioASP, http://www.bioasp.nl) and BioMed Grid Portal (Biomed Grid Portal BI, http://bmg.bii.a-star.edu.sg) also use a single system to combine databases and analyses, but are not provided with strong dynamic and interactive visualization. Grids and Web-based portals are typically built around a single database and support a limited number of tools that are not so flexible and extensible. Centralization of only analysis and visualization components of the workflow forms another system called "super applications," (e.g. J-Express Pro (MolMine)). Super applications do not support the process of data acquisition and do not support plug-ins for new analysis or visualization components. PathPort employs ToolBus as a client-side interconnect allowing data from different sources to be connected with a number of visualization tools on the desktop, while running Web-services on the server-side to provide data access and analysis. Therefore, PathPort is not limited to a single server or virtual organization for data access and analysis.

The BioWidgets (Fischer, et al., 1999) and ISYS (Siepel, et al., 2001) are two component-based programs freely available for non-commercial uses. The BioWidgets toolkit is a collection of Java Beans used for development of graphical application and/or applets in the genomics domain. BioWidgets wires components directly to one another to provide top-level controls. The components in BioWidgets require customized, tight integration resulting in less flexibility and more development time. ISYS is a decentralized platform with a component-based approach for the integration of heterogeneous bioinformatics software tools and databases. ISYS is also written in Java and supports web-based resources including databases, analysis tools or websites. In many cases, ISYS allows separate components to share data and exchange services by implementing a system architecture similar to that in PathPort. Like ToolBus in PathPort, the ClientBus in ISYS is the central part responsible for all communication between components.

The main difference between ISYS and PathPort is that PathPort uses standard XML-based Web Services making the system easily scalable and extensible. Current emphasis is on the use of XML technologies. PathPort XML-based data formats allow for easy data communication between different machines and data transformation involving varied formats. PathPort does not require data reformatting between different tools for the same set of data and specific data can also be selected and drag-dropped into several analysis tools. Furthermore, different data can be grouped and compared for common or differential annotations. Entire PathPort work session(s) can be saved for subsequent analysis or alternatively, the data can be compressed and electronically sent to collaborators allowing them to view and/or continue analysis of the data.

It is noted that the ToolBus architecture is a universal integration system and can be used to integrate any data and analysis resources via a XML-based web service system. ToolBus is currently used in PathPort for interoperable pathogen-oriented data and tools. PathPort's first year efforts have focused on genomic data and related information such as DNA sequences, genome annotations, genome comparisons, and phylogenetic relationships. Current and future focus will be on microarray oligo design and analysis, proteomics data, metabolomics data, geospatial data, and other environmental data. Corresponding tools will also be developed as needed to integrate and analyze these additional data types allowing relationships between these data to be discovered and shared by users. Distributed biological data management and analysis necessitates the development of standards for molecular and cellular data, especially in terms of communication standards.

The I3C is committed to bringing public and private members together to develop XML-based standards for life science data (Interoperable Informatics Infrastructure Consortium; http://www.i3c.org). VBI is part of a collaboration to improve the current DAS system for standardized distribution of genome sequence information. The development and use of PathInfo and ProNet XML DTDs at VBI also provides possible solutions for standardization of distributed pathogen background information and molecular interactions in systems other than bioterrorism such as plant pathogens or food-borne pathogens. There is also a critical need to develop annotation standards and knowledge representation ontologies for molecular and cellular data. Current efforts include the Gene Ontology (GO) Consortium (Consortium GO, 2001), the COG database (Tatusov, et al., 2000), the InterPro database (Apweiler, et al., 2001) and the Sequence Ontology database (Sequence Ontology Database, http://song.sourceforge.net/).

PathPort/ToolBus was designed as a scalable, extendable, framework for integration of highly curated pathogen related data and genome analysis/visualization tools. We have used a limited set of tools to demonstrate integrative uses of PathPort/ToolBus. The PathPort tools used in the *Brucella* spp. genome annotation and comparison are available as stand alone applications or super applications that can be found on the web, however, the major advantage of PathPort/ToolBus is its provision of these tools as a single, interoperable resource. Pathogen-related information is scattered throughout the literature and at numerous websites (for example, http://www.bt.cdc.gov; http://www.e-bioterrorism.com; http://www.bioterrorism.uab.edu; http://www.ci.berkeley.ca.us/publichealth/bioterrorism/bioterrorismmain.html). The information available at these sites contains information of limited use to researchers working on fundamental problems in infectious diseases. None of these sources provide information about the molecular interactions that take place during pathogenesis and they lack detailed information about the newest diagnostic methods. The pathogen information provided through PathPort's PathInfo and ProNet databases provides a current, expertly curated compilation of data regarding the organism, its associated disease, standard and novel diagnostic methods, and when available, information and visualization tools for the mechanism of pathogenesis. In summary, we know of no other single system with similar data or interoperable analysis and visualization tools such as those provided by PathPort/ToolBus, to aid in infectious diseases research and development.

ToolBus can be freely downloaded at the VBI PathPort official website (http://www.vbi.vt.edu/pathport) under non-commercial license. An evaluation license is available for potential commercial users. The ToolBus download also includes a variety of ToolBus visualizers, most of which are made available under the terms of the Lesser GNU Public License (LGPL) (http://www.gnu.org/copyleft/lesser.html). It is noted that some visualizers may have more restrictive licenses (e.g., the WebWindow visualization plugin).

ACKNOWLEDGEMENTS. The PathPort Project, which resulted in the creation of ToolBus, was supported by Department of Defense grant (DAAD 13-02-C-0018) to Dr. Bruno Sobral. We are grateful to our corporate partners, IBM and Sun Microsystems, for their support of this work via an IBM Shared University Research Award and a Sun Center of Excellence in Bioinformatics to Dr. Bruno Sobral. We are grateful to our collaborators at Soldier Biological and Chemical Command (Edgewood, MD), in particular, Dr. Jay Valdes and his team, for feedback throughout the software development process. Contributions of the PathPort biologists and IT scientists are gratefully acknowledged.

## 5. REFERENCES

Achard, F., Vayessix, G., and Barillot, E., 2001, XML, bioinformatics and data integration. *Bioinformatics*. **17**:115.

Apweiler, R., Attwood, T.K., Bairoch, A., et al., 2001, The InterPro database, an integrated documentation resource for protein families, domains and functional sites. *Nucleic Acids Res.* **29**: 37.

Bateman, A., Birney, E., Cerruti, D., Etwiller, L., Eddy, S.R., Griffiths-Jones, S., Howe, K.L., Marshall, M., and Sonnhammer, E.L.L., 2002, The Pfam protein families database. Nucleic Acids Res. **30**: 276.

BioASP, http://www.bioasp.nl.

Biomed Grid Portal BI, http://bmg.bii.a-star.edu.sg.

Consortium, G.O., 2001, Creating the gene ontology resource: design and implementation. *Genome Res.* **11**: 1425.

Conway, T., and Schoolnik, G.K., 2003, Microarray expression profiling: capturing a genome-wide portrait of the transcriptome. *Mol. Microbiol.* **47**: 879.

Delcher, A.L., Harmon, D., Kasif, S., White, O., and Salzberg, S.L., 1999, Improved microbial gene identification with GLIMMER. *Nucleic Acids Res.* **27**: 4636.

Delcher, A.L., Philippy, A., Carlton, J., and Salzberg, S.L., 2002, Fast algorithms for large-scale genome alignment and comparison. *Nucleic Acids Res.* **30**: 2478.

Dowell, R.D., Jokerst, R.M., Day, A., Eddy, S.R., and Stein, L., 2001, The Distributed Annotation System. *BMC.Bioinformatics* **2**: 7.

Eckart, J.D., and Sobral, B.W.S., 2004, *A life scientist's gateway to distributed data management and computing: thePathport/ToolBus framework.* OMICS 2003: InPress.

Eckart, J.D., Sobral, B.W.S., Laubenbacher, R., and Mendes, P., 2002, *The role of bioinformatics in toxicogenomics and proteomics.* Proceedings from NATO Advanced Workshop on Toxicogenomics and Proteomics. October 16–20, 2002, Prague, Czech Republic.

Fischer, S., Crabtree, J., Brunk, B., Gibson, M., and Overton, G.C., 1999, bioWidgets: data interaction components for genomics. *Bioinformatics* **15**: 837.

Foster, I., and Kesselman, C. 1998, The Globus project: A status report. Proc Heterogenous Computing Workshop, pp. 4–18.

Gardner, T., 2001, An Introduction to Web Services. Ariadne, Issue 29. 02-October-2001. http://www.ariadne.ac.uk

Grimshaw, A., 2003, al. e. in: *Grid Computing: Making the global infrastructure a reality*, edited by F. Berman, ed., Wiley.

Interoperable Informatics Infrastructure Consortium; http://www.i3c.org.

MolMine, J.P., http://www.molmine.com/frameset/frm_jexpress.htm.

NCBI genomes. http://www.ncbi.nlm.nih.gov/Genomes/index.html

Rouillard, J.M., Herbert, C.J., and Zuker, M., 2002, OligoArray: genome-scale oligonucleotide design for microarrays. *Bioinformatics* **18**: 486.

Rozen, S., and Skaletsky, H., 2000, Primer3 on the WWW for general users and for biologist programmers. Methods *Mol. Biol.* **132**: 365.

Siepel, A., Farmer, A., Tolopko, A., et al., 2001, ISYS: a decentralized, component-based approach to the integration of heterogeneous bioinformatics resources. *Bioinformatics* **17**: 83.

Sequence Ontology Database. http://song.sourceforge.net/.

Tatusov, R.L., Galperin, M.Y., Natale, D.A., and Koonin, E.V., 2000, The COG database: a tool for genome-scale analysis of protein functions and evolution. *Nucleic Acids Res.* **28**: 33.

*Chapter 14*

# Alignment of Wheat and Rice Structural Genomics Resources

Daryl J. Somers, Sylvie Cloutier, and Travis Banks

## 1. INTRODUCTION

Whole plant genome sequences hold substantial promise as the next genetic resource to change agricultural sciences. Recently, the draft rice sequence was released with efforts from Syngenta (Goeff et al. 2002) and the International Rice Genome Project (http://rgp.dna.affrc.go.jp/IRGSP/) (Yu et al. 2002). Monsanto and the Clemson University Genome Institute (CUGI) have also played a large role in both sequencing and rice genome organization. As a cereal geneticist, this is an exciting time, as we now consider how to use the rice genome sequence to understand genome organization and gene regulation in wheat, barley and maize.

Researchers have examined wheat for decades in an attempt to locate genes and QTLs on genetic maps that give us insight into which chromosomal regions control the traits important for grain production, processing and nutrition. These same QTLs have been examined further to devise and/or validate molecular breeding strategies to accelerate development of improved wheat varieties through marker-assisted selection (MAS). Since rice and wheat are similar cereal grain crops, the colinearity of genes and gene sequence are sufficient to align the chromosomes of each species (Devos and Gale, 2000).

The logical next step in wheat genomics is locating wheat QTLs onto rice chromosomes and ultimately associating wheat QTLs with rice genome sequence

**Daryl J. Somers, Sylvie Cloutier, and Travis Banks**    Agriculture and Agri-Food Canada Cereal Research Centre, Winnipeg, Manitoba, Canada R3T-2M9.

*Genome Exploitation: Data Mining the Genome*, edited by J. Perry Gustafson, Randy Shoemaker, and John W. Snape.
Springer Science + Business Media, New York, 2005.

and candidate genes. As this type of alignment matures and is refined, identification of genes controlling traits in wheat will improve, leading to careful genetic engineering of wheat with a higher end use value or consistent production.

Over the last four years, a vast amount of wheat EST sequencing has been completed and released to the public domain. In March 2003, there were >415,000 wheat EST sequences (http://www.ncbi.nlm.nih.gov/). The wheat research community is fortunate that much of this sequencing was coordinated and that there was deliberately a large number of wheat accessions that were used as template for the cDNA library construction and EST sequencing. This leads immediately to mining of the databases for allelic variation and single nucleotide polymorphisms (SNPs). SNPs represent the most basic form of DNA polymorphism and they are useful in all the traditional molecular marker applications, from mapping to genetic diversity studies.

Bread wheat (*Triticum aestivum*) has the genome constitution AABBDD and provides a novel challenge in SNP discovery and analysis since it is a polyploid species, carrying three progenitor genomes derived from *T. monococcum* (AA), *T. speltoides* (BB) and *T. tauschii* (DD). ESTs in wheat will originate from all three genomes and in order for SNP diagnostic tests to be useful as genetic markers they should have a strong degree of locus specificity. Therefore, SNP mining strategies and detection platforms need to account for this added layer of complexity.

Given the size of the wheat genome, estimated to be 1.6 billion bases per 1C (Arumuganathan and Earle, 1991) distributed in 21 linkage groups, the efforts to align wheat and rice genomics resources will be time consuming and expensive. It will be critical to share the workload across the international wheat community in a coordinated fashion in order to ensure our accuracy and success.

## 2. PRIMARY STRUCTURAL GENOMICS RESOURCES

A short description of the basic wheat and rice structural genomics resources is needed in order to follow the interrelationships of the wheat and rice genomes. The primary goal of linking the structural genomics resources is to discover candidate rice genome sequence spanning a wheat chromosome interval for the purpose of gene discovery.

### 2.1. Genetic Maps

Wheat and rice have been mapped genetically with DNA-based markers for over 15 years. Much of the earlier work was based on RFLP markers, and eventually, cDNA probes were used on both maps. The cDNA probes, which represent genes, could be cross hybridized to both species, and thus the colinear relationship between wheat and rice, and in fact, many other cereals, was developed (Devos and Gale, 2000; Bennetzen, 2002; Sarma et al., 2000). Microsatellites represent a profound change in marker technology. They are typically locus specific, codominant, polymorphic and PCR-based. These attributes make microsatellites highly amenable to plant breeding and high throughput structural genomics studies. Microsatellite-based maps are now high density and together with cDNA RFLP maps, provide a good alignment of wheat and rice maps with good resolution.

## 2.2. QTL Mapping

Once genetic maps were available, QTL analysis followed and genes controlling end use quality, stress resistance, growth habit and disease resistance as examples were located onto chromosomes. The QTL analysis is a statistical association of alleles identified along the length of a chromosome with a phenotype. Often, comparative mapping studies show genes from syntenic chromosome regions of wheat and rice, controlling similar traits. QTL analysis in wheat provides the targets of possible map-based cloning efforts. For example, the genes controlling vernalization, grain hardness, and photoperiod sensitivity have been examined extensively. Similarly, genes controlling leaf rust (Feuillet et al., 2001) and Fusarium resistance (Somers et al., 2003; Anderson et al., 2001) are good targets as these genes may elucidate the mechanisms of disease resistance.

## 2.3. EST Sequences

There are >415,000 wheat EST sequences in the public domain as of March 2003. This genomics resource was developed by coordinated efforts including the International Triticeae EST Consortium (ITEC), the NSF funded EST sequencing project in Albany, CA (http://wheat.pw.usda.gov/NSF/), and a substantial public release of sequences into NCBI from Dupont in late 2002. The EST sequences are a rich source of information with respect to allelic variation and gene expression influenced by both biotic and abiotic stress. The ESTs are useful for discovering SNPs, as there are now bioinformatics approaches to mining the EST database for this type of allelic variation (Marth et al., 1999; Yuan et al., 2001; Somers et al., 2003). The current collection of >415,000 EST has now been assembled into contigs (http://wheat.pw.usda.gov/ITMI/2002/WheatSNP.html), which gives insights into the number of genes coded in the wheat genome and present in the cDNA libraries. The statistics of the Dec 2002 EST assembly show 39,813 contigs, 50,116 singletons. The ESTs in this collection have been used for chromosome bin mapping using a unique set of deletion lines derived from Chinese Spring (CS) wheat described below.

## 2.4. Chinese Spring Deletion Lines

The polyploid nature of wheat permits the development of unique cytogenetic stocks such as nullisomic, tetrasomic, substitution and addition lines. Having multiple copies of genes on different genomes maintains the viability and fertility of these types of lines. A novel set of CS lines developed by Endo and Gill (1996) at Kansas State University, contain terminal disomic deletions. The breakpoints on the chromosomes are random and the breakpoints are identified in a physical sense by measuring chromosome sizes at metaphase under a microscope. A set of $>100$ of these lines have been selected that have single breakpoints each and the collection of lines represents breakpoints across the genome. This set of lines has been used in Southern analysis with ESTs as probes to map the physical location of EST sequences to physical bins in the wheat genome (http://wheat.pw.usda.gov/NSF/). In total, $>6,200$ Loci have been mapped into chromosome bins. The CS deletion lines are being distributed worldwide and are being used to map other markers such as microsatellites into these bins, which helps align the genetic and physical maps of wheat (Somers et al., PAG XI, 2003, http://www.intl-pag.org/).

## 2.5. Large Insert Libraries

Large insert libraries are an important resource for physical mapping, map-based cloning, genomic sequencing as well as genome evolution and comparative genomic studies (Anderson et al., 2003, Dubcovsky et al., 2001, Tranquilli et al., 1999). Yeast Artificial Chromosome (YAC), Bacterial Artificial Chromosome (BAC) and cosmid libraries have all been used for these purposes. In the last decade, a large number of BAC libraries were constructed because of their large size, their plasmid nature and therefore the ease of manipulation and stability of the clones. Many rice BAC libraries were constructed (Chen et al., 2002; Tao et al., 1994; Tao et al., 2002; Yang et al., 1997; Zhang et al., 1996). Also several wheat BAC libraries were constructed in diploid, tetraploid and hexaploid wheat (Lijavetzky et al., 1999, Liu et al., 2000, Moullet et al., 1999). A list of Triticeae large insert libraries can be found at http://agronomy.ucdavis.edu/Dubcovsky/BAC-library/ITMIbac/ITMIBAC.htm.

In rice, BAC libraries were fingerprinted and BAC clones were assembled into contigs (Presting et al., 2001). The minimum tilling path was constructed and sequencing of the entire rice genome is near completion. Hexaploid wheat has a genome size 40 times greater than rice (Arumuganathan and Earle, 1991). So far, the sequencing of its entire genome has not been contemplated. However, global fingerprinting of the D-genome is underway and would provide the basis for genomic sequencing of a diploid progenitor. Regions comprising important agronomic traits such as the HMW glutenins and the hardness locus have been sequenced in *T.tauschii* (Anderson et al., 2003, Tranquilli et al., 1999).

Because of the size of wheat BAC libraries, clones cannot be individually screened. Many strategies and tools have been designed to overcome this difficulty. High-density filters where clones are double-spotted in a predetermined pattern using high-precision robotic equipment have been generated for many available wheat BAC libraries. These filters can be hybridized with probes to identify clones of interest. A drawback of this method is that the larger the size of the library, the larger the number of filters required. A $4 \times 4$ pattern is the most commonly used density. To print a 500,000 clone library at this density, a total of 27 filters would be required. One way to cope with the large size of the wheat BAC libraries is to increase the density on the filters. Chalhoub (personal comm) reported that sensitivity was still good at a density of $6 \times 6$. At that density, a 500,000 clone

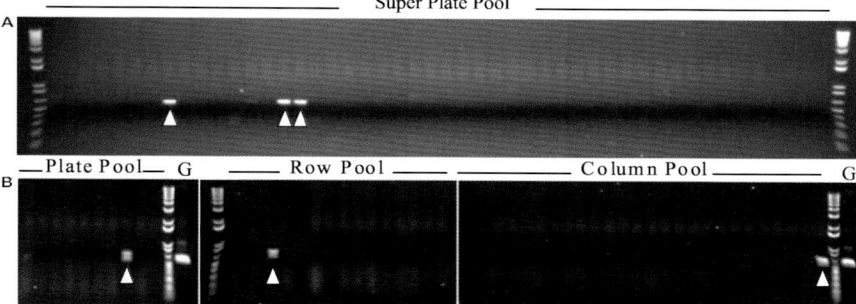

**FIGURE 1.** Screening of the BAC pools using a pair of primers designed to target the grain softness protein (GSP) gene. A. Panel of 48 of the 171 super plate pools showing three positive pools ($\triangle$). B. Screening of the ten corresponding plate pools, 16 row pools and 24 column pools indicating the address of the positive BAC clone ($\triangle$) which is plate 8, row 4 and column P (column 24). G = Glenlea.

library can be printed onto 12 filters only. Considering that current BAC libraries of diploid wheat comprise 150,000 to 300,000 clones, tetraploid wheat comprise 500,000 clones and hexaploid wheat comprise 600,000 to 1.2 million clones, increased density on the BAC filters will not only be advantageous but will be a prerequisite to perform such screening of the libraries. High-density filters of the hexaploid wheat BAC library of Glenlea were generated using a $5 \times 5$ gridding pattern. Sensitivity at that density was still excellent. However, filter hybridization still remains labor intensive and the screening of BAC filters for the identification of clones of interest is likely going to be restricted to specific cases.

Pooling of BAC clones is an efficient method to screen large number of clones by PCR (Nilmalgoda et al., 2003). We devised a BAC pooling strategy that allows the screening of the 650,000 clones of the Glenlea library in a total of 221 PCR reactions performed in two steps. The first step consists of 171 PCR reactions of the super plate pools (SPPs). The second round of PCR is 50 reactions representing the 10 plate pools, 16 row pools and 24 column pools for each positive SPP. These 221 reactions allow the identification of the positive clone amongst the entire library (Figure 1). This method is simple and efficient providing that the clone(s) targeted can be screened by PCR.

## 2.6. Rice Genome Sequence

A draft sequence of two subspecies of rice (*Oryza sativa* L. spp japonica, *Oryza sativa* L. spp indica) was completed and released by Goff et al. (2002) and Yu et al. (2002) respectively. The japonica genome is estimated to be 430 MB in length containing 32,000–50,000 genes and the genome sequence covers 93% of the total. Wheat, barley and maize ESTs were compared via TBLASTN to the japonica genome sequence and 98% of the ESTs found a homologue in the rice sequence. Similarly, the indica genome sequence is estimated to be 466 MB containing 46,000–55,000 genes and 92% of the genome has been sequenced. Sequence comparisons were made to *Arabidopsis* in this case, and not to other cereal genomes. The rice sequence from IRGSP is available at http://btn.genomics.org.cn/rice.

Assuming that wheat chromosome regions can be aligned with syntenic regions of the rice genome, then it should be possible to predict genes and gene sequences in certain regions of wheat chromosomes. The most anticipated aspect of this type of orthologous genome alignment is searching for candidate genes for wheat QTLs using the rice genome sequence. Sequencing the wheat genome, or portions of it, is feasible, particularly if gene rich islands can be identified among wheat BAC clones, to restrict the amount of sequencing. But, undoubtedly, there will be regions of the wheat genome not amenable to sequencing and the rice genome sequence will be utilized. We have developed a software tool, based on all the publically available information and bioinformatics resources described above. This effort is described below and investigates whether wheat/rice synteny is robust enough to facilitate wheat candidate gene discovery within QTL regions.

## 3. SNPs IN WHEAT ESTs

Single nucleotide polymorphisms (SNPs) are regarded as the next type of molecular marker most useful for genetic studies. Allelic variation at such loci is restricted to four alleles, and the vast majority of SNP loci in wheat have just two alleles. So, the allelic variation for SNPs is minimal, but 1) the immense number of SNPs present 2) the ability to mine SNPs from the EST database and 3) high throughput detection platforms, make SNPs a very attractive marker type in genetic studies. Mining SNPs from the EST database is a bioinformatics and PCR-based challenge (Marth et al., 1997; Somers et al., 2003). In most cases, the EST sequence quality data is not publicly available and assembly of wheat ESTs creates contigs with copies of ESTs derived from homoeologous and paralogous loci. Molecular marker diagnostic tests should be locus specific and polyploidy presents the challenge of developing these types of tests. Fortunately, the assembly of the EST sequences aligns EST sequences from homoeologous and paralogous loci. This provides the opportunity to design PCR primers capable of amplifying with locus specificity.

### 3.1. Nested PCR Analysis of SNPs

A technique such as nested PCR can be used, where primers flanking the SNP amplify the locus from one genome and a third primer interrogates the SNP on the amplicon (Figure 2). Detection platforms for SNPs are varied. Our lab has chosen to use nested PCR followed by single base extension using the SNAPshot $^{TM}$ kit (Perkin Elmer). This assay is desirable, because it begins with genome specific PCR then uses dye-deoxy-terminators to colour-code the allele present in the DNA sample. The products of the single-base extension are separated by electrophoresis on an ABI3100 capillary electrophoresis genotyper. The presence of SNPs in

**FIGURE 2.** Cartoon of nested PCR in hexaploid wheat. Three primers are used, the flanking primer 3′ ends anneal to nucleotide positions (△) that are specific to the A, B or D genome. The internal primer 3′ end anneals to a SNP (▲) that differs among wheat accessions.

wheat ESTs (Somers et al. 2003) and the EST assembly in the public domain (http://wheat.pw.usda.gov/ITMI/2002/WheatSNP.html) can be used in a practical sense to design diagnostic tests for known functional genes.

## 3.2. Development of Genome-Specific SNPs

Three copies of a given gene are often found in hexaploid wheat, each one corresponding to one of the genomes. These homoeologous copies can differ greatly due to evolution but can sometimes share a high level of homology. The large number of wheat ESTs available often permits the identification of the three genome-specific copies based on SNPs. The following example, based on an NBS-LRR sequence, illustrates this fact. The 2,836 nucleotide sequence of clone GLN164A08 was BLASTed against the contig assemblies. The best match corresponded to contig-6605.1 (Dec02 assembly). SNPs can be easily visualized (Figure 3). EST

```
                    *      760       *       780       *       800
contig6605 : ATAAGCGCGGAAACGTCATGGAGAGATCCATTGTTGATCC*TCTT*AGACT*GTTGTTCA :  795
CSPBJ22873 : ATAAGCGCGGAAACGTCATGGAGAGATCCATTGTTGATCC*TCTT*AG----------- :  664
CSPBJ22401 : ATAAGCGCGGAAACGTCATGGAGAGATCCATTGTTGATCC*TCTT*AGACT*GTTGTTCA :  385
SUMBM13655 : ATAAGCGCGGAAACGTCATGGAGAGATCCATGGTCGATCC*TCTT*AGACT*GTTGTTCA :  272
U20CA63685 : ATAAGCGCGGAAACGTCATGGAGAGATCCATTGTTGATCC*TCTT*AGACT*GTTGTTCA :  243
STECA65855 :-ATAAGCGCGGAAACGTCATGGAGANATCCATTGTTAATCCCTCTTTAGACTTGTTGTTCA :   60
MERAL82541 :----------------------AGATCCATGGTCGATCC*TCTT*AGACT*GTTGTTCA :   34
U19CA64611 :------------------------------CCTCTT*AGACT*GTTGTTCA :   19
CSPBE40684 :------------------------------------------------------------ :    -
CSPBF20214 :------------------------------------------------------------ :    -
7

GLN164A08  :GATAAGCGCGGAAACGTCATGGAGAGATCCATTGTTGATCC*TCTT*AGACT*GTTGTTCA :  641

                    *      1880      *       1900      *       1920
contig6605 : TCCTCGAGCTGTGAGAGAGACCACACATCATCAAGTATAACAAGAACTGGTCCTCGGTCCC : 1905
CSPBJ22873 :------------------------------------------------------------ :    -
CSPBJ22401 :------------------------------------------------------------ :    -
SUMBM13655 :------------------------------------------------------------ :    -
U20CA63685 :------------------------------------------------------------ :    -
STECA65855 :------------------------------------------------------------ :    -
MERAL82541 :------------------------------------------------------------ :    -
U19CA64611 :------------------------------------------------------------ :    -
CSPBE40684 : TCCTCGAGCTGTGAGAGAGACCACACATCATCAAGTATAACAAGAACTGGTCCTCGGTCCC :  326
CSPBF20214 : TCCTCGAGCTGTGAGAGAGACCACACATCATCAAGTATAACAAGGACTGGTCCTCGGTCCC :   88
GLN164A08  : TCCTCGAGCTGTGAGAGAGGCCACACATCATCAAGTATAACAAGGACTGGTCCTCGGTCCC : 1751
```

**FIGURE 3.** Partial alignment of wheat ESTs from contig 6605.1 with NBS-LRR clone GLN164A08. Genome-specific SNPs are in bold face. Primer sequences are underlined. Variety nomenclature is a three-letter code representing the accession or library followed by the GenBank accession number. CSP is Chinese Spring; SUM is Sumai 3; STE is Stephens; MER is Mercia; U19 and U20 are from two different library of unknown accession; Contig6605 is the consensus sequence; GLN164A08 is the NBS-LRR clone from Glenlea wheat. Consensus sequence base positions are on top. Base position for each EST is on the right.

5A 5B 5D

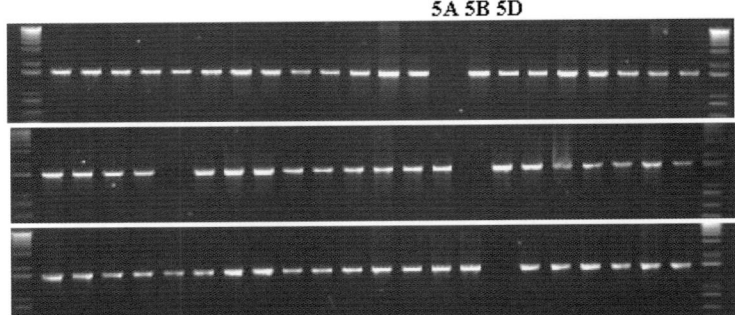

**FIGURE 4.** PCR amplification of Chinese Spring and its 21 nulli-tetra lines using genome-specific SNP primers designed for the NBS-LRR gene located on homoeologous group 5 chromosome. The 5A-specific amplicon was obtained with primer pair F1128T/R2280GC; the 5B-specific amplicon used primer combination F1128C/R2280GC; and the 5D-specific amplicon was generated with F1128C/R2281TT. Primer sequences are as follows: F1128C: 5′-AGGGACCGAGGACCAGTC-3′; F1128T: 5′-AGGGACCGAGGACCAGTT-3′; R2281TT: 5′-CGTCATGGAGAGATCCATTGTT-3′; R2280GC: 5′-GTCATGGAGAGATCCATGGTC-3′.

containing SNPs can often be grouped into two or three clusters within a contig. Primers for genome-specific SNPs of this sequence were designed and tested using the CS nulli-tetra series (Figure 4). The amplification obtained confirmed that homoeologous copies of clone GLN164A08 were located on chromosome 5A, 5B and 5D. Each copy differed from one another by a small number of conserved single nucleotide polymorphism.

## 4. WHEAT/RICE VIRTUAL MAPPING (WRVM)

The availability of the draft genome sequence of rice offers the opportunity to take this information and apply it to the study of other grasses. Rice has a small genome of approximately 430 MB with 12 chromosomes and recent efforts by the International Rice Genome Sequencing Project (IRGSP) have placed the vast majority of the genome sequence into the public domain (http://rgp.dna.affrc.go.jp/IRGSP/). Initially, Clemson University Genomics Institiute (CUGI) determined a genome-tiling path of rice BAC clones and these clones were taken by various IRGSP partners, sequenced and released. The tiling paths of BACs were then assembled into contigs and aligned against a rice genetic map to give an ordering of the contigs along the rice genome. The genomic information by itself is of limited value, however organizations such as The Institute for Genomic Research (TIGR) have analyzed this data to mine for secondary sequence information such as introns, exons, and repeat sequences (ftp://ftp.tigr.org/pub/data/Eukaryotic_Projects/o_sativa/annotation_dbs/). The automated process used by TIGR to predict gene-coding regions has identified over 72,000 potential genes in rice. TIGR acknowledges this as an overestimation due to duplication of some genes since BACs overlap. As rice is an established model species, we can anticipate that other grass species may have a similar number of genes, many of which will be highly related.

The large size of the wheat genome means that an effort to sequence the entire genome is not in the immediate future. This leaves the wheat research community without a 'directory' to locate the position of wheat gene sequences on the wheat genome. To overcome this limitation we can use the information extracted from rice and combine it with other *Triticeae* resources such as the EST bin-mapping project by the USDA, in-house mapping of genetic markers to the bin map, and the large amount of publicly available EST sequence data to develop a gross physical map of the wheat genome.

The syntenic relationship between rice and wheat is the basis for the Wheat Rice Virtual Map (WRVM). Though separated by millions of years of evolution, both rice and wheat contain genomic structure from the ancestral parent from which both species evolved. Each of the bread wheat genomes is extremely homologous to one another and each of these genomes can be viewed as a shuffled and diverged version of the wheat/rice ancestral parent. The rice genome can also be considered a diverged and re-ordered wheat/rice ancestral parent (Devos and Gale, 2000). Despite insertion, deletion and inversion events in the genomes of rice and wheat there is still underlying conservation of the genes and gene-order between the species. The wheat virtual physical map takes advantage of the conservation of the genes and their ordering along the genome to predict the location of wheat genes on the bin map based on the experimentally derived EST bin locations and the gene neighbours of the wheat EST rice homologs.

The 6,218 ESTs placed into the wheat bin by Southern hybridization (http://wheat.pw.usda.gov/NSF/progress_mapping.html) were compared against the set of predicted rice genes from TIGR using the BLASTN program (Altschul, et al., 1997). Any wheat sequences having greater than or equal to 100 nucleotides identical to a rice gene with an alignment expect value of $<=$ 1e-50 were considered to be homologous sequences. Of the 6,218 ESTs, 2,221 wheat sequences were found to have rice homologs. These rice genes will be referred to as 'anchors' from here on. The contig sequences from the wheat EST assembly project represent a set of wheat genes. The 419,347 wheat EST sequences from Genbank are represented by 89,929 sequences from the assembly project (http://wheat.pw.usda.gov/ITMI/2002/WheatSNP.html), each of which was compared against the putative rice gene collection from TIGR using BLASTN. A total of 62,824 of these found a rice 'hit' with the default BLASTN cutoff values. The rice anchor sequences are used to 'bring' the surrounding rice genomic sequence to the same wheat bin as the wheat homolog of the rice anchor sequence. Any of the wheat contigs from the Wheat EST assembly project that have a best match to one of the rice genes in the surrounding genomic sequence brought to the bin are considered to be located in that bin. In this way the wheat ESTs are given a physical location and a virtual map of the wheat genome is constructed (Figure 5). The user is then able to search for the location of an EST by specifying a Genbank accession number or sequence and optionally a wheat marker that has been placed on the bin map.

RICE                                                                              WHEAT

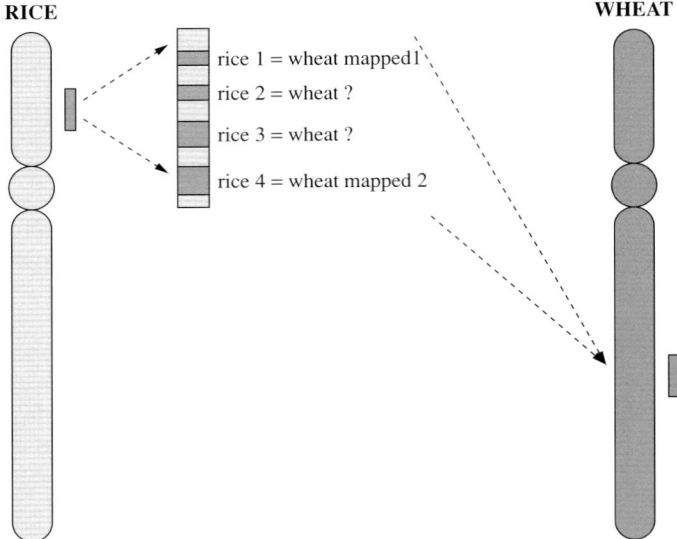

**FIGURE 5.** Mapping of rice segments to wheat bins. A portion of rice genomic DNA (solid rectangle) contains 4 rice genes. Two of the genes are homologous to bin mapped ESTs (rice 1 and rice 4). These are the rice anchor sequences. Both of the mapped wheat ESTs are in the same bin on the long arm of the wheat chromosome and therefore bring in the section of rice genomic DNA to that bin. If any of the 89,929 wheat representative sequences have a best match to rice gene 2 or 3 they are virtually mapped to the same bin as the rice section.

For a variety of reasons (discussed below) an EST can be found at a number of locations on the virtual map. To further delimit the results returned, the user can specify a marker that has been bin mapped so only EST positions that are in the same bin as the marker are presented. This allows the user to take a gene of interest and see if it is located near a marker for a trait, such as disease resistance. The markers were not mapped to the same resolution as the USDA ESTs and therefore the mapped location of an EST can differ from the predicted position, though the bin of the predicted gene will be wholly contained within the bin having the marker.

This approach to constructing a wheat virtual map has some limitations. First, many of the genes that are of interest to researchers may not be present in rice and therefore the wheat gene will not be on the map. Also, functional homologues of the wheat genes may exist in rice but there may be so much sequence divergence that the software does not recognize their relationship. Another limitation of the map is that bin-mapped wheat ESTs have been assigned to many different bins on different chromosomes. This may represent members of the same gene family that are detected by the same Southern probe or gene duplications that do not represent syntenic regions. In either case the map will assume synteny between that wheat region and rice and the software will construct potentially false map locations.

## 5. IMPLEMENTATION OF 'WRVM'

Eleven wheat ESTs were mapped via SNPs on reference populations and then bin mapped to vaidate the WRVM. When a sequence is found on the map, the justification for the positioning of that sequence is given as the number of anchor sequences that have brought the rice homolog of the wheat EST into that bin. The greater the number of anchor sequences the stronger the likelihood that the regions are truly syntenic and the mapping is correct. The number of anchors returned for a sequence is dependent on the size of the region around the virtually mapped EST brought to the bin. The larger the region to search for anchors the stronger the mapping can be, but it also increases the number of false mappings presented to the user.

Of the 11 EST sequences placed on the bin map, 10 were found on the virtual map (Table 1). Six out of 10 of these, BM140481, BM134420, BE497718, BF202619, BE499017 and BE418437 are believed to have been mapped correctly. For example, BM140481 was placed by Southern hybridization onto the long arm of chromosome 1A. The virtual map placed this EST into bin 1AL3-0.61-1.00, a bin comprising the last 39% of the long arm of chromosome 1A. BM134420 was mapped to the long arm of chromosome 2B and was placed by the software on the long arm of chromosome 2D in bin 2DL3-0.48-0.76. When delimiting by a marker the virtual map assumes that homoeologous chromosomes are identical so the long arm of 2B is identical to the long arms of 2A and 2D. BE499017 was found by the software in the experimentally derived bin, as was BE418437. The incorrect mapping of the other 4 sequences, particularly those sequences with multiple anchors, may be due to genome duplication where the SNP marker was not specific for the duplicated gene but the region contained enough homology to allow for interaction with a Southern probe. BQ619858 is a good example of this scenario as the EST had 6 anchors to a bin other than the one it was mapped to. It is also possible the EST bin mapping data is incorrect. False map locations can also be due to single gene duplications or the virtual map assuming relationships between rice and wheat sequences that do not exists. In either case the result is the software displaying incorrect syntenic relationships leading to false positions and an overall increase in the complexity of the map. Given the current structural genomics resources integrated in this project, the software does succeed at directing the user to the most likely wheat/rice alignment.

## 6. SUMMARY

A time line for large scale sequencing of the wheat genome or gene rich regions and accurate annotation of the sequence is difficult to predict. This sort of resource would lead directly to gene discovery in wheat for many disease resistance and seed quality traits. As a bridge to this scheme, the rice genome sequence indirectly provides insight into genes and DNA sequence in the vicinity of wheat QTLs.

## Table 1
## Virtual Positions of Bin Mapped Wheat ESTs Based on Homology to Predicted Rice Open Reading Frames.

[a] Neighbour BAC+-2

| Genbank | Exp. Bin | Best match | # of Anchors | Delimiting Marker | Best match incl. marker | # of Anchors |
|---|---|---|---|---|---|---|
| BM140481 | C-1AL-1.00 | 4DS2-0.82-1.00 | 2 | Gwm357 | 1AL3-0.61-1.00 | 1 |
| 10BQ619858 | C-1AL-1.00 | C-3AL3-0.42 | 2 | Gwm357 | None | 0 |
| BM134420 | C-2BL-1.00 | 2DL3-0.48-0.76 | 7 | Gwm55 | 2DL3-0.48-0.76 | 7 |
| BE497718 | 2DS-0.47-1.00 | 2BL6-0.89-1.00 | 3 | Gwm102 | None | 0 |
| BF482680 | C-4BS-0.57 | 4DS2-0.82-1.00 | 2 | Gwm113 | None | 0 |
| BG909876 | C-5AS-0.40 | 2AS5-0.78-1.00 | 1 | Barc180 | None | 0 |
| BF202619 | 6BL-0.40-1.00 | C-6AL4-0.55 | 2 | Wmc182 | None | 0 |
| BE499017 | 7BS-0.27-1.00 | 4AL4-0.80-1.00 | 2 | Wmc76 | 7BS1-0.27-1.00 | 1 |
| BQ239140 | 7BL-0.78-1.00 | | | NO HIT | | |
| BE418437 | 7DS-0.61-1.00 | 4AL5-0.66-0.80 | 3 | Barc70 | 7DS4-0.61-1.00 | 2 |

[b] Neighbour BAC+-10000

| Genbank | Exp. Bin | Best match | # of Anchors | Delimiting Marker | Best match incl. marker | # of Anchors |
|---|---|---|---|---|---|---|
| BM140481 | C-1AL-1.00 | 4DS2-0.82-1.00 | 2 | Gwm357 | 1AL3-0.61-1.00 | 1 |
| BQ619858 | C-1AL-1.00 | C-3AL3-0.42 | 6 | Gwm357 | None | 0 |
| BM134420 | C-2BL-1.00 | 2DL3-0.48-0.76 | 10 | Gwm55 | 2DL3-0.48-0.76 | 10 |
| BE497718 | 2DS-0.47-1.00 | 2BL6-0.89-1.00 | 3 | Gwm102 | None | 0 |
| BF482680 | C-4BS-0.57 | 4DS2-0.82-1.00 | 2 | Gwm113 | None | 0 |
| BG909876 | C-5AS-0.40 | 2AS5-0.78-1.00 | 1 | Barc180 | None | 0 |
| BF202619 | 6BL-0.40-1.00 | C-6AL4-0.55 | 6 | Wmc182 | None | 0 |
| BE499017 | 7BS-0.27-1.00 | 4AL5-0.66-0.80 | 4 | Wmc76 | 7AS5-0.59-0.89 | 3 |
| BQ239140 | 7BL-0.78-1.00 | | | NO HIT | | |
| BE418437 | 7DS-0.61-1.00 | 4AL5-0.66-0.80 | 6 | Barc70 | 7DS4-0.61-1.00 | 5 |

[a] (**Neighbour BAC+-2**)—The accession number of the wheat EST is given in the first column with the experimentally determined bin map position in the second. The best match is the location of the EST with the highest number of anchor sequences. Column 5 has the name of the marker that was used to limit the search results for any bins wholly contained within the bin containing the marker. Each of the chosen markers are located in the same experimentally determined position as the EST. The final column is the number of anchor sequences that justify the prediction of the EST being in the same bin region as the marker. Anchor sequences are taken from two rice BACs upstream of the virtual hit through to two rice BACs downstream.
[b] (**Neighbour BAC+- 10**)—The same as above except that the anchor sequences are taken from the 10 BACs upstream through 10 BACs downstream of the virtual hit on the rice genomic sequence. Increasing the genomic area around the hit to search increases the number of anchors for $3/4$ of the correctly mapped ESTs.

To build the structural genomic relationship between wheat and rice, many components are integrated including wheat microsatellite maps, a wheat EST bin map, wheat microsatellite bin map, a contig assembly of >415,000 wheat ESTs, ordered rice.

BACs, and the draft rice genome sequence. All of this is tied together in a piece of software written by Travis Banks which can be upgraded as more quantity and accurate information is developed.

When these genomics resources are improved and the interrelationships are refined, the prospect of discovering candidate wheat ESTs or genes in QTL intervals using the rice genome sequence will improve. We are currently investigating wheat chromosome regions controlling FHB resistance, wheat midge resistance, and will examine regions controlling seed quality characteristics in the future. Ultimately, wheat gene discovery is refined to a wheat BAC contig where BAC sequencing will be required. The 3.1X hexaploid wheat BAC library at AAFC is providing a direct flow of research from wheat-fine mapping (based on rice sequences) to BAC clone isolation and sequencing.

## 7. REFERENCES

Altschul, S.F., Madden, T.L., Schaffer, A.A., Zhang, J., Zhang, Z., Miller, W., and Lipman, D.J., 1997, "Gapped BLAST and PSI-BLAST: a new generation of protein database search programs". *Nucleic Acids Res.* **25**:3389–3402.

Anderson, D., Rausch, C., Moullet, O., and Lagudah, S., 2003, The wheat D-genome HMW-glutenin locus: BAC sequencing, gene distribution, and retrotransposon clusters. *Functional Integrative Genomics* **3**:56–68.

Anderson, J.A., Stack, R.W., Liu, S., Waldron, B.L., Fjeld, A.D., Coyne, C., Moreno-Sevilla, B., Mitchell Fetch, J., Song, Q.J., Cregan, P.B., and Frohberg, R.C., 2001, DNA markers for *Fusarium* head blight resistance QTL in two wheat populations. *Theor Appl Genet.* **102**:1164–1168.

Arumuganathan, K., and Earle, E.D., 1991, Estimation of nuclear DNA content of plants by flow cytometry. *Plant Mol Biol Rep* **9**:229–241.

Bennetzen, J., 2002, Opening the door to comparative plant biology. *Science* **296**:60–63.

Chen, M., Presting, G., Barbazuk, W.B., Goicoechea, J.L., Blackmon, B., Fang, G., Kim, H., Frisch, D., Yu, Y., Sun, S., Higingbottom, S., Phimphilai, J., Phimphilai, D., Thurmond, S., Gaudette, B., Li, P., Liu, J., Hatfield, J., Main, D., Farrar, K., Henderson, C., Barnett, L., Costa, R., Williams, B., Walser, S., Atkins, M., Hall, C., Budiman, M.A., Tomkins, J.P., Luo, M., Bancroft, I., Salse, J., Regad, F., Mohapatra, T., Singh, N.K., Tyagi, A.K., Soderlund, C., Dean, R.A., and Wing, R.A., 2002, An Integrated Physical and Genetic Map of the Rice Genome. *The Plant Cell* **14**:537–545.

Devos, K.M., and Gale, M.D., 2000, Genome relationships: the grass model in current research. *The Plant Cell* **12**:637–646.

Dubcovsky, J., Ramakrishna, W., SanMiguel, P.J., Busso, C.S., Yan, L., Shiloff, B.A., and Bennetzen, J.L., 2001, Comparative sequence analysis of colinear barley and rice bacterial artificial chromosomes. *Plant Physiol* **125**:1342–1353.

Endo, T.R., and Gill, B.S., 1996, The deletion stocks of common wheat. J. of Heredity **87**(4):295–307.

Feuillet, C., Penger, A., Gellner, K., Mast, A., and Keller, B., 2001, Molecular evolution of receptor-l Like kinase genes in hexaploid wheat. Independent evolution of orthologs after polyploidization and mechanisms of local rearrangements at paralogous loci. *Plant Physiol* **125**:1304–1313.

Goff, S.A., Ricke, D., Lan, T.-H., Presting, G., Wang, R., Dunn, M., Glazebrook, J., Sesions, A., Oeller, P., Varma, H., Hadley, D., Hutchison, D., Martin, C., Katagiri, F., Lange, B.M., Moughamer, T., Xia, Y., Budworth, P., Zhong, J., Miguel, T., Paszkowski, U., Zhang, S., Colbert, M., Sun, W.-L., Chen, L., Copper, B., Park, S., Wood, T.C., Mao, L., Quail, P., Wing, R., Dean, R., Yu, Y., Zharkikh, A., Shen, R., Sahasrabudhe, S., Thomas, A., Cannings, R., Gutin, A., Pruss, D., Reid, J., Tavtigian, S., Mitchell, R.M., Bhatnagar, S., Adey, N., Rubano, T., Tusneem, N., Robinson, R., Feldhaus, J., Macalma, T., Oliphant, A., and Briggs, S., 2002, A draft sequence of the rice genome (*Oryza sativa* L. ssp. Japonica). *Science* **296**:92–100.

Lijavetzky, D., Muzzi, G., Wicker, T., Keller, B., Wing, R.A., and Dubcovsky, J., 1999, Construction and characterization of a bacterial artificial chromosome (BAC) library for the A genome of wheat. *Genome* **42**:1176–1182.

Liu, Y.G., Nagaki, K., Fujita, M., Kawaura, K., Uozumi, M., and Ogihara, Y., 2000, Development of an efficient maintenance and screening system fro large-insert genomic DNA libraries of hexaploid wheat in a transformation-competent artificial chromosome (TAC) vector. *Plant J* **23**:687–695.

Marth, G.T., Korf, I., Yandell, M.T., Yeh, R.T., Gu, A., Zakeri, H., Stitzel, N.O., Hillier, L., Kwok, P.-Y., and Gish, W.R., 1999, A general approach to single-nucleotide polymorphism discovery. *Nature Genetics* **23**:452–456.

Moullet, O., Zhang, H.B., and Lagudah, E.S., 1999, Construction and characterisation of a large DNA insert library from the D genome of wheat. *Theor Appl Genet* **99**:305–313.

Nilmalgoda, S.N., Cloutier, S., and Walichnowski, A.Z., 2003, Construction and characterization of a bacterial artificial chromosome library of hexaploid wheat (*Triticum aestivum* L.) and identification of clones containing a sequence from high molecular weight glutenin gene Bx7. *Genome* (submitted).

Presting, G.G., Budiman, M.A., Wood, T., Yu, Y., Kim, H.R., Goicoechea, J.L., Fang, E., Blackman, B., Jiang, J., Woo, S.S., Dean, R.A., Frisch, D., and Wing, R.A., 2001, A framework for sequencing the rice genome. *Novartis Found Symp* **236**:13–24.

Sarma, R.N., Fish, L., Gill, B.S., and Snape, J.W., 2000, Physical characterization of the homoeologous group 5 chromosomes of wheat in terms of rice linkage blocks, and physical mapping of some important genes. *Genome* **43**:191–198.

Tao, Q., Wang, A., and Zhang, H.B., 2002, One large-insert plant-transformation-competent BIBAC library and three BAC libraries of Japonica rice for genome research in rice and other grasses. *Theor Appl Genet* **105**:1058–1066.

Tao, Q.Z., Zhao, H.Y., Qiu, L.F., and Hong, G.F., 1994, Construction of a full bacterial artificial chromosome (BAC) library of *Oryza sativa* genome. *Cell Res* **4**:127–133.

Tranquilli, G., Lijavetzky, D., Muzzi, G., and Dubcovsky, J., 1999, Genetic and physical characterization of grain texture-related loci in diploid wheat. *Mol gen Genet* **262**:846–850.

Yang, D., Parco, A., Nandi, S., Subudhi, P., Zhu, Y., Wang, G., and Huang, N., 1997, Construction of a bacterial artificial chromosome (BAC)library and identification of overlapping BAC clones with chromosome 4-specific RFLP markers in rice. *Theor Appl Genet* **95**:1147–1154.

Yu, J., Hu, S., Wang, J., Wong, G.K.S., Li, S., Liu, B., Deng, Y., Dai, L., Zhou, Y., Zhang, X., Cao, M., Liu, J., Sun, J., Tang, J., Chen, Y., Huang, X., Lin, W., Ye, C., Tong, W., Cong, L., Geng, J., Han, Y., Li, L., Li W., Hu, G., Huang, X., Li, W., Li, J., Liu, Z., Li, L., Liu, J. Qi, Q., Liu, J., Li., L., Li, T., Wang, X., Lu, H., Wu, T., Zhu, M., Ni, P., Han, H., Dong, W., Ren, X., Feng, X., Cui, P., Li, X., Wang, H., Xu, X., Zhai, W., Xu., J., Zhang, K., Zheng, X., Dong, J., Zeng, W., Tao, L., Ye, J., Tan, J., Ren, X., Chen, X., He, J., Liu, D., Tian, W., Tian, C., Xia., H., Bao, Q., Li G., Gao, H., Cao, T., Wang, J., Zhao, W., Li, P., Chen, W., Wang, X., Zhang, Y., Hu, J., Wang, J., Liu, S., Yang, J., Zhang, G., Xiong, Y., Li, Z., Mao, L., Zhou, C., Zhu, Z., Chen, R., Hao, B., Zheng, W., Chen, S., Guo, W., Li, G., Liu, S., Tao, M., Wang, J., Zhu, L., Yuan, L., and Yang, H., 2002, A draft sequence of the rice genome (*Oryza sativa* L. spp. Indica). *Science* **296**:79:92.

Yuan, Q., Quackenbush, J., Sulfana, R., Pertea, M., Salzberg, S.L., and Buell, C.R., 2001, Bice bioinformatics. Analysis of rice sequence data and leveraging the data to other plant species. *Plant Physiol.* **125**:1166–1174.

Zhang, H.B., Choi, S., Woo, S.S., Li, Z., and Wing, R.A., 1996, Construction and characterization of two rice bacterial artificial chromosome libraries from the parents of a permanent recombinant inbred mapping population. *Mol Breed* **2**:11:24.

*Chapter 15*

# Computational Identification of Legume-Specific Genes

Michelle A. Graham, Kevin A. T. Silverstein, Steven B. Cannon, and Kathryn A. VandenBosch

## 1. IDENTIFICATION OF SEQUENCES OF INTEREST FROM PUBLIC DATABASES

One of the biggest challenges facing biologists today is the sheer amount of sequence information available. There are more than 1.4 million sequences in the GenBank nonredundant database and more than 16.3 million sequences in dbEST (National Center for Biotechnology Information [NCBI], April, 2003). To access sequences of interest from within GenBank, two main options are available. The first is to use the BLAST algorithm to query a previously identified sequence of interest against any of the GenBank sequence databases (Altschul et al., 1997). This will identify sequences with similarity to the query sequence. Results can be delimited by species of interest and by degree of similarity to the query sequence. The second option for accessing sequence data from GenBank is to query the database using keyword searches. In this case, searches can be limited by species of interest and/or by keywords anywhere in the sequence annotation. Both methods have distinct advantages and disadvantages. BLAST can be used to identify all sequences that have sequence similarity to a specific nitric oxide synthase, a calmodulin binding protein. However, a keyword search would be necessary to identify all other classes of proteins that may also bind calmodulin.

**Michelle A. Graham, Kevin A. T. Silverstein, Steven B. Cannon and Kathryn A. VandenBosch**
Department of Plant Biology, University of Minnesota, St. Paul, Minnesota, 55108.

*Genome Exploitation: Data Mining the Genome*, edited by J. Perry Gustafson, Randy Shoemaker, and John W. Snape.
Springer Science + Business Media, New York, 2005.

Once sequences of interest have been identified, other problems can arise. Many sequences lack informative annotation. These include 'putative' or 'hypothetical' sequences identified from genomic DNA that are predicted to encode a gene. Other sequences may be incorrectly annotated or may be annotated by their expression pattern in a particular species. A second problem is sequence redundancy. In the case of *Medicago truncatula*, there are 443 EST contigs whose ESTs are represented more than fifty times in dbEST (TIGR Medicago truncatula Gene Index 6; http://www.tigr.org/tdb/tgi/mtgi). The results of BLAST and keyword searches often fail to answer basic biological questions: Where is the gene expressed? How related are orthologous genes? Is the mRNA alternatively spliced? Is the gene part of a gene family? Given these problems, it can be difficult to frame biological questions that will yield clear answers. The focus of this paper will be to use sequence databases to ask important questions about legume biology.

## 2. IMPORTANCE OF LEGUMES

Grain legume species such as peas, beans, and lentils account for over 33% of human dietary protein nitrogen and needs worldwide (Vance et al., 2000). Other legumes, such as alfalfa, clover, and barrel medics (*M. truncatula*) are widely used as animal fodder. Legumes are important sources of protein, oil, mineral nutrients, and nutritionally important natural products, such as flavanoids. One important feature of legumes is their interaction with *Rhizobium* bacteria. Ongoing communication between plant and *Rhizobium* is essential in development and maintenance of the nodule, a plant organ found only in legumes and a small number of related taxa (Doyle and Luckow, 2003). Through this symbiotic interaction, legumes supply usable nitrogen to both natural and agricultural ecosystems. Legumes also form symbiotic associations with mycorrhizal fungi, which aid in mineral acquisition from the soil. While these interactions are beneficial to the plant, many fungal and bacterial pathogens also exist which can have a large impact on crop yields. The many uses of legumes and the variety of symbiotic and pathogenic interactions found provide numerous targets for functional genomics research.

## 3. DATABASES OF EXPRESSED SEQUENCE TAGS

Currently, expressed sequence tags (ESTs) are being collected from three legume species. There are 308,582, 181,444, and 36,210 EST sequences available for soybean, *M. truncatula*, and *Lotus japonicus*, respectively (NCBI, April, 2003). Once the individual EST sequences are generated, they are assembled into contigs, which represent the minimally redundant set of a species' expressed genes. Different research groups have used different algorithms to assemble the ESTs into contigs (VandenBosch and Stacey, 2003). The Institute for Genomic Research (TIGR) has clustered the ESTs from a variety of species, including the three legume species, into minimally redundant unigene sets (TCs; Quackenbush et al., 2001).

## 3.1. Data Mining the TIGR Gene Indices

Using the TIGR Gene Indices, researchers can query ESTs by gene product name, using an ortholog from the Eukaryotic Gene Orthologs (EGO; Lee et al., 2002) database, or by expression pattern. In addition, 'Electric northern blots' can be performed using Boolean operators. This allows the user to select contigs shared between libraries or contigs that are unique to a specific experimental condition. Using Boolean searches, Federova et al. (2002) identified 340 TCs with nodule-specific expression patterns from the TIGR *M. truncatula* Gene Index 4. Many of the genes identified shared sequence similarities with nodulins, plantacyanin, calmodulins, purine permease, and an embryo-specific protein. In addition, a family of 114 TCs with weak similarity to the early nodulin ENOD3 was also identified, which will be discussed in detail below. One of the drawbacks of this approach is that genes expressed at low levels may not be accurately reported. To determine the validity of using *in silico* analyses to predict gene expression, the expression patterns of 91 TCs were examined by northern blot. TCs composed of more than five ESTs did indeed have nodule enhanced expression patterns.

## 3.2. Data Mining the *M. truncatula* MtDB Database

In *M. truncatula*, sequencing has focused on ESTs representing different stages of development, different organs, and responses to microbial interactions such as symbiotic nodulation, colonization or pathogen infection. One goal of the NSF-funded *M. truncatula* Consortium was to build a database, MtDB, (Lamblin et al., 2003; http://www.medicago.org/MtDB) for Medicago EST sequences that was easily accessible, flexible, and could be used to answer basic biological questions of interest to researchers. The Center for Computational Genomics and Bioinformatics (CCGB, University of Minnesota) have separately clustered and assembled the Medicago ESTs into a unigene set, housed within MtDB. The database will also contain additional genetic, genomic, and biological information about *M. truncatula*. Like the TIGR Gene Indices, MtDB can be searched for contigs made up of combinations of ESTs from particular libraries. In addition, users can limit their searches by the BLAST results of the contigs. For example, to search for legume-specific genes involved in nodulation, a user can easily search for contigs expressed exclusively in nodules that have BLAST hits to legume species but not nonlegume species. In the mycorrhizal root libraries, some of the ESTs might not be of Medicago origin since the endophytic fungal symbiont might have also contributed to the polyA RNA pool used for library construction. A user could search for contigs from mycorrhizal root libraries with BLAST homology to fungi but not plants to identify these sequences. A similar approach can be utilized to identify ESTs corresponding to pathogens in the pathogen inoculated libraries.

One important feature of MtDB is the ability to check the quality of the contigs. A user can compare equivalent contigs identified from TIGR and MtDB. Since the contigs were assembled using different algorithms (Cap4 and Phrap), some contig differences may exist. The results of the comparison include ESTs that are shared between the two contigs and ESTs that are unique to each contig. The equivalent contigs query can also be used to identify contigs within MtDB that are likely part of a gene family or represent alternative splice products. Finally, users can view a graphical representation of an MtDB contig to determine if the assembly is correct. Each of the four nucleotides is represented in a different color and sequence quality is indicated by the height of the base. With this contig representation, problems in contig assembly can often be easily recognized.

## 4. IDENTIFICATION OF LEGUME-SPECIFIC GENES

With the database tools in place, we could begin to ask questions about legume biology. As mentioned previously, the formation of *Rhizobium*-induced nitrogen-fixing nodules is a phenomenon unique to legumes and a small number of related species (Doyle and Luckow, 2003). In addition, legumes can synthesize isoflavanoids not found in other plant species. While many of the genes involved in these processes were likely adopted from other pathways shared between many different plant species, some of the genes may be unique to legumes or may have diverged so much within legumes that they appear unique. The goal of this project was to identify sequences from *M. truncatula*, soybean, and *L. japonicus* that appeared to be legume-specific. Preliminary results from this study were reported in an overview of the Medicago genome project (VandenBosch and Stacey, 2003). As a working definition, legume-specific genes have no sequence homology, below a specified threshold, to any publicly available sequences of nonlegumes. Note that putative legume-specific genes may ultimately have nonlegume homologs from species that aren't represented in the sequence databases.

### 4.1. Using BLAST to Identify Legume-Specific Genes

To identify legume-specific genes, the NCBI BLASTN and TBLASTX programs were used to compare the TIGR Medicago, soybean (*Glycine max* and *Glycine soja*), and Lotus TCs against the TIGR maize (ZmGI), tomato (LeGI), rice (OsGI), and *Arabidopsis* (AtGI) Gene Indices (Table 1; Quackenbush et al., 2001; http://www.tigr.org/tdb/tgi). The Gene Indices included both TC and singleton EST sequences. By using a limited dataset of nonlegume species, we hoped to quickly and efficiently eliminate as many TCs as possible that had homology to sequences from nonlegumes. If a legume TC had BLAST homology to any sequence from one of the four nonlegume species with an E-value more significant than $10^{-4}$, it was not considered legume-specific and was removed from the list of putative legume-specific genes. Using this two-step BLAST approach, we were

**Table 1**
**Identification of Legume-Specific Genes Using the BLAST Programs to Compare Legume and Nonlegume Sequences.**

| Legume Data | | Glycine max/soja | Medicago truncatula | Lotus |
|---|---|---|---|---|
| Legume TCs | Medicago, Soybean and Lotus TCs[a] | 24,750 | 17,243 | 3,790 |
| BLAST Program | Nonlegume data set | Legume-specific Glycine max/soja TCs remaining | Legume-specific Medicago truncatula TCs remaining | Legume-specific Lotus japonicus TCs remaining |
| BLASTN | TIGR AtGI, LeGI, OsGI, and ZmGI (142,492 sequences)[b] | 7,938 | 5,309 | 886 |
| TBLASTX | TIGR AtGI, LeGI, OsGI, and ZmGI (142,492 sequences) | 2,412 | 1,417 | 144 |
| BLASTX | GenBank nonredundant database (1,335,905 sequences) | 2,230 | 1,267 | 136 |
| TBLASTX | Remaining TIGR Plant GIs (325,748 sequences)[c] | 2,101 | 1,141 | 110 |
| TBLASTX | TIGR Arabidopsis thaliana and Oryza sativa genomes | 2,081 | 1,128 | 106 |
| Tera-BLASTN | EST_others (6,985,891 sequences) | 2,020 | 1,046 | 103 |
| Tera-TBLASTX | EST_others (6,985,891 sequences) | 1,997 | 1,028 | 101 |

[a] Soybean (GmGI, v9), Medicago truncatula (MtGI, v6) and Lotus japonicus (LjGI, v1)
[b] Arabidopsis thaliana (AtGI, v9), Tomato (LeGI, v9), Oryza sativa (OsGI, v10), and Zea mays (ZmGI, v10)
[c] Barley (HvGI, v5), Chlamydomonas reinhardtii (ChrGI, v2), cotton (CGI, v4), grape (VvGI, v1), ice plant (McGI, v3), lettuce (LsGI, v1), Pinus spp. (PGI, v1), potato (StGI, v6), rye (RyeGI, v2), Sorghum bicolor (SbGI, v4), sunflower (HaGI, v1), and wheat (TaGI, v6)

able to eliminate approximately 90% of TCs from Medicago, soybean and Lotus that did not appear to be legume-specific.

With a smaller subset of putative legume-specific TCs remaining, we could use more stringent and computationally intensive searches to identify homologous sequences from nonlegume species. At each of the following steps, either the number of sequences in the nonlegume dataset was increased or the stringency of the BLAST search was changed from BLASTN to BLASTX or TBLASTX. In all cases, legume TCs with homology to nonlegume sequences with an E-value more significant than $10^{-4}$ were not considered legume-specific.

The next step was to take the remaining 3,973 legume-specific TCs and compare them to the GenBank protein database (nr) using BLASTX. BLASTX translates a legume TC in all six reading frames and compares the results to a protein database. An additional 340 TCs were identified with homology to nonlegume sequences, leaving 3,633 legume-specific TCs. Next, the remaining legume-specific TCs were compared to the remaining nonlegume plant Gene Indices available at TIGR. This included the Gene Indices for barley, *Chlamydomonas reinhardtii*, cotton, grape, ice plant, lettuce, *Pinus* spp., potato, rye, sorghum bicolor, sunflower, and wheat. Comparisons were made using TBLASTX, which translates the query and target sequences in all six reading frames and compares all 36 combinations to find possible homology. Another 283 legume TCs with homology to nonlegume sequences were identified.

All of the BLAST searches described to this point involved nonlegume sequences that most likely represent real genes. This includes expressed genes, such as ESTs, genes that have been identified experimentally, and also genes that have been predicted from genomic sequences. Gene prediction programs use different algorithms to identify coding and noncoding sequences. It is unlikely that any gene prediction software could identify all potential genes. Therefore, it was necessary to search the remaining legume-specific TCs against the available genome sequences of *Arabidopsis thaliana* (The Arabidopsis Genome Initiative, 2000; Bevan et al., 2001) and *Oryza sativa* (Goff et al., 2002; Yuan et al., 2003). Using TBLASTX, the 3,352 remaining legume-specific TCs were compared against the rice and *Arabidopsis* genomes. Unlike previous searches, the sequences were filtered prior to BLAST to remove low complexity sequences. This resulted in the removal of an additional 37 legume TCs.

For the last steps, it was important to compare the remaining legume-specific TCs with the largest dataset, NCBI's EST-others, which contains all EST sequences except for those from human and mouse. With more than 16 million sequences, this was by far the most computationally intensive search. If the same machine was used for this step as had been used previously, a TBLASTX search of the remaining 3,315 legume-specific TCs would have taken more than six weeks. Therefore, we used the DeCypher® Bioinformatics Accelerator running the Tera-BLAST™ hardware accelerated version of BLAST (TimeLogic, Crystal Bay, NV) which was housed in CCGB at the University of Minnesota. Using this system, we were able to run Tera-BLASTN and Tera-TBLASTX against EST-others in less than three hours. After successive BLAST searches, 1,028,

1,997, and 101 legume-specific TCs remained from Medicago, soybean, and Lotus, respectively.

## 4.2. Are the Identified Legume-Specific TCs Really Legume-Specific?

One area of concern was that some of the legume-specific TCs might be too short to give informative BLAST hits. Hits could also be missed if a TC's sequence corresponded to the untranslated portion of the gene and not to the protein-encoding portion of the gene. Since the untranslated portions of the gene do not encode a protein, they are less likely to be conserved among species. To get around this problem, the remaining 3,126 putative legume-specific TCs were compared to the original set of legume TCs using TBLASTX at different stringency levels (Figure 1). If a legume-specific TC had a BLAST hit only to itself or to another legume-specific TC, it was retained as legume-specific. However, if a legume-specific TC had a BLAST hit to a TC with homology to nonlegumes, it was given lower priority. Based on the small inflection at $1.00E^{-20}$, this value was chosen as the cutoff for further restricting the legume-specific TCs. Using this cutoff, 2,525 legume-specific TCs remained. Of these, less than 5% had significant sequence homology to legume sequences in the GenBank nonredundant database. Sequence homologies include: late nodulins (*Vicia faba*), hypothetical proteins (*Galega orientalis*, *Glycine max*, and *Cicer arietinum*), nodule-specific protein and seed albumin PA1 (*Pisum sativum*), MtN1 and MtN17 (*M. truncatula*), leginsulin and 1,3-$\beta$-glucanase (*Glycine max*) and TrPRP2 (*Trifolium replens*).

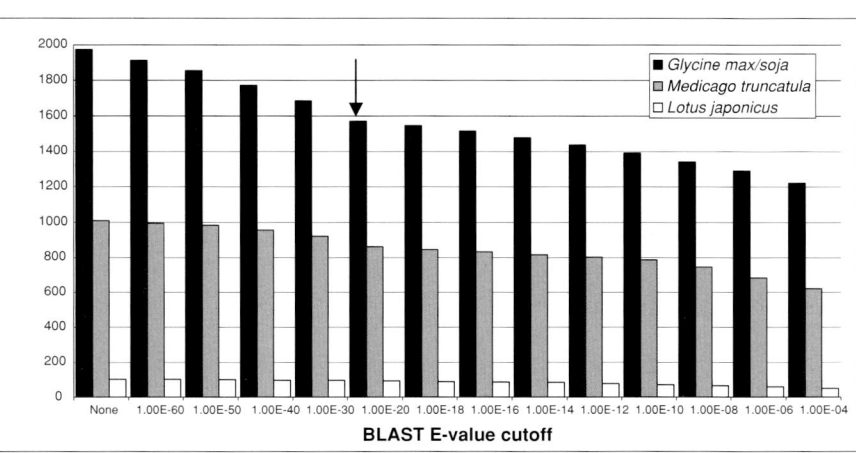

**FIGURE 1.** The effect of different E-values when using BLASTX to compare putative legume-specific TCs against all legume TCs. Putative legume-specific TCs with BLAST hits to TCs with homology to nonlegume sequences at a particular E-value are removed. Stringency of the cutoff increases from left to right across the x-axis. The arrow indicates a small inflection in the graph at $1.00E^{-20}$. This cutoff was used to further restrict the number of legume-specific TCs.

## 5. HOW CAN A GENE'S FUNCTION BE DETERMINED WHEN BLAST FAILS TO FIND A HOMOLOG?

### 5.1. Single Linkage Clustering

In order to identify legume-specific TCs that were members of a gene family or that had homologs among the other examined legume species, single linkage clustering was performed. TBLASTX was used to compare all legume-specific TCs against themselves using an E-value cutoff of $10^{-6}$. If the BLAST reports of two TCs had a least one TC in common, the TCs were combined into a group. For example, if the BLAST report of TC1 overlapped with BLAST report of TC2, and the BLAST report of TC2 overlapped with the BLAST report of TC3, all three TCs and the TCs included in their BLAST reports would be put in a single group. Using this technique, 124 different groups representing 439 TCs were identified (Table 2).

Surprisingly, of the 2,525 TCs that were clustered, only 31 groups were identified that contained TCs from both Medicago and Glycine. Since sequences from both species were well represented, this result was unexpected. In an effort to explain the lack of homology between Medicago and Glycine, the tissue source of ESTs within legume-specific TC was examined in each of the species (Figure 2). In Medicago, 69% of ESTs from legume-specific TCs corresponded to nodule or root cDNA libraries. In contrast, only 16% of ESTs from legume-specific TCs in

Table 2
**Identification of Homologous Legume-Specific TCs using Single Linkage Clustering.**

| Groups of Homologous Genes Identified Across Species | Number of Groups Identified |
|---|---|
| *Medicago truncatula* and *Lotus japonicus* | 2 |
| *Medicago truncatula* and *Gycine max/soja* | 30 |
| *Medicago truncatula*, *Glycine max/soja*, and *Lotus japonicus* | 1 |
| Homologous groups identified within *Medicago truncatula* | |
| 2 TCs per group | 27 |
| 3 TCs per group | 4 |
| 4 TCs per group | 7 |
| 5 TCs per group | 1 |
| 7 TCs per group | 1 |
| 10 TCs per group | 1 |
| 125 TCs per group | 1 |
| Homologous groups identified within *Glycine max/soja* | |
| 2 TCs per group | 55 |
| 3 TCs per group | 4 |
| 4 TCs per group | 1 |
| 5 TCs per group | 1 |
| 15 TCs per group | 1 |
| Homologous groups identified within *Lotus japonicus* | |
| 2 TCs per group | 2 |

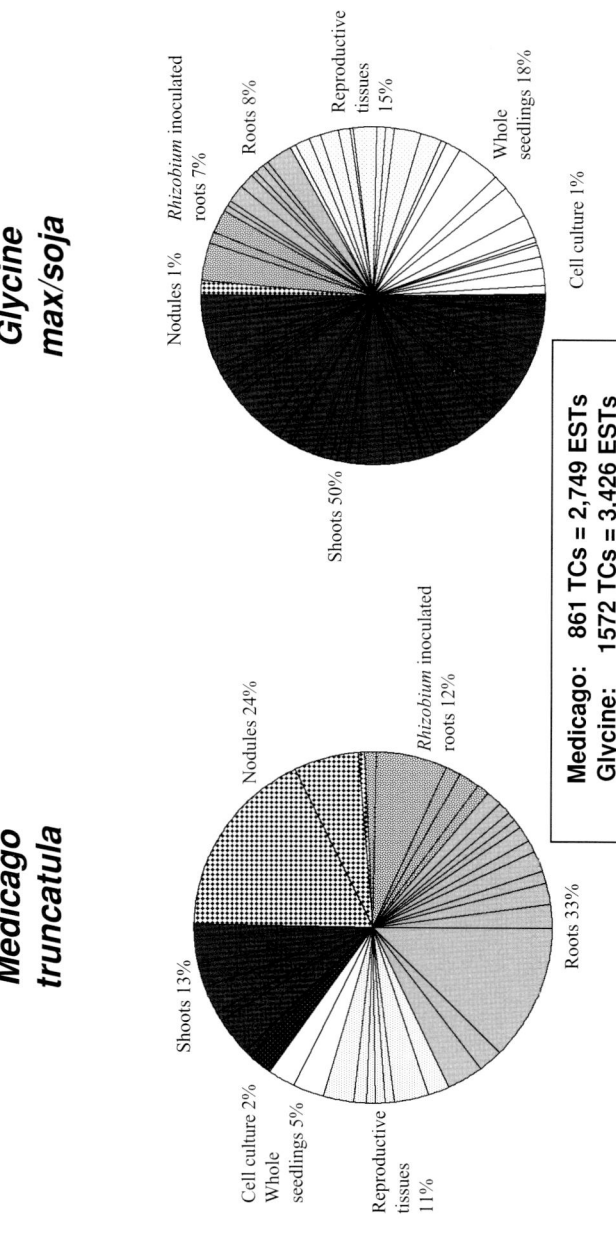

**FIGURE 2.** EST distribution of legume-specific TCs from *Medicago truncatula* and *Glycine max/soja*. Individual pie slices represent unique cDNA libraries from which ESTs were sequenced. Libraries corresponding to similar tissues have been color coded and labeled. In Medicago, 69% of ESTs from the legume-specific TCs correspond to roots; Rhizobium inoculated roots, and nodules. In soybean, 68% of EST from legume-specific TCs correspond to shoots and whole seedlings.

soybean came from nodule or root libraries. Instead, more than 50% of the legume-specific soybean ESTs came from shoot libraries. A similar result was observed when the distribution of ESTs from all nonhomologous Medicago and soybean TCs was examined (data not shown). This suggests that the lack of homology between the two species is due to tissue bias in the EST sequencing projects rather than species differences. Using single linkage clustering, a gene family of 125 Medicago TCs was identified. Some of the TCs within this group had weak ($>10^{-5}$) similarity to nodule-specific cDNAs GoAT-Ldd1 (*Galega orientalis*; Kaijalainen et al., 2002), cysteine cluster proteins (CCPs,*Vicia faba*, Frühling et al., 2000) or ENOD3 (*Pisum sativum*, Scheres et al., 1990). Each of these TCs had a small open reading frame of approximately 70 amino acids. The 5′ end of the predicted protein contained a cleavable signal peptide, while the remainder of the protein contained two cysteine clusters (Figure 3). Each cluster was composed of two cysteine residues, almost always separated by five amino acids. When the TCs were used as query sequences for TBLASTX analysis of the Medicago singletons, more than 150 singleton sequences were identified that were members of this large gene family. Together, all the members of the family account for 6% of the ESTs identified from the mature and senescent nodule libraries (GVN and GVSN). Sequence alignment of the CCPs reveals the high degree of sequence divergence between members of the gene family (Figure 3). This group was simultaneously identified by Fedorova et al. (2002) because of its nodule-specific expression pattern and also by Mergaert et al. (2003) who used BLAST to query the Medicago ESTs for sequences with homology to cysteine cluster proteins identified by Györgyey et al. (2000).

A second interesting group bore a lot of similarity to the nodule-specific CCPs. This group contained ten TCs from Medicago and two TCs from soybean. When the cut-off for single linkage clustering was lowered to $10^{-4}$, this group would cluster with the nodule-specific CCPs. Like the nodule-specific CCPs, members of this group had a cleavable signal peptide and two cysteine clusters. However, none of the members in this group had significant BLAST hits. In addition, all members of the group appeared to have seed-specific expression patterns.

## 5.2. Motif Searching

Many of the groups of legume-specific genes we identified had little sequence homology to characterized genes or gene products. Hence, we sought alternative methods to hypothesize the functions of these families. Motif-searching can be useful in characterizing genes when BLAST fails to find a homolog. Unlike BLAST, motif-searching algorithms weigh highly conserved residues more heavily, allowing a multilevel consensus. In addition, they require no minimum word-length match (a consecutive string of matched bases) and the more sequences that are built into the motif, the better it becomes at finding a match in a query sequence. One option for motif-searching was to characterize legume-specific TCs using a collection of known motifs. Programs such as PROSITE (Falquet et al., 2002), InterProScan (Mulder et al., 2003), PRINTS (Attwood et al., 2003), PFAM (Bateman et al.,

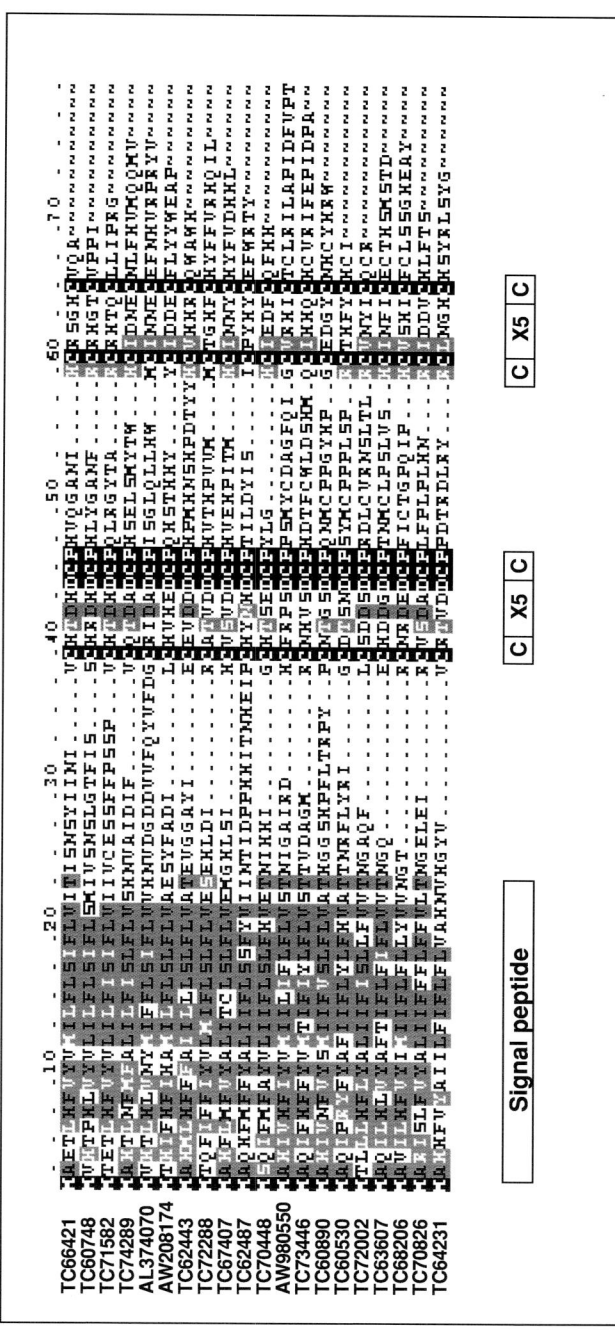

**FIGURE 3.** Multiple sequence alignment of *Medicago truncatula* cysteine cluster proteins (CCPs). Only a subset of the CCP family is shown. Singletons sequences are designated by their GenBank accession nembers. Identical residues in the alignment are shown as black boxes with white text. Conserved residues are shown as grey boxes with black text. Similar residues are shown as grey boxes with white text. Below the alignment, a diagram is used to identify conserved structures.

2002), SMART (Letunic et al., 2002), TIGRFAMs (Haft et al., 2003), and PANAL (Silverstein et al., 2000) can be used to search a query sequence for conserved motifs that have been identified from other proteins. A second option was to determine whether some motifs that were conserved among a group of legume-specific genes bore any similarity to parts of previously characterized proteins. If there was a common theme among the hit proteins, this would provide a hint of possible function. The following procedure was used to model, identify, and refine motifs shared within each group of legume-specific TCs having three or more sequence members:

1. TBLASTX of legume-specific TCs was used to identify singleton sequences that could also belong to each group.
2. Each TC and singleton within a group was translated in all six frames.
3. The entire set of translated sequences was modeled as an ungapped position-specific scoring matrix (PSSM) using the MEME program (Bailey and Elkan, 1994). This program uses an expectation maximum algorithm to identify optimal-width motifs among the protein sequences. In this step, most of the incorrect frames from the translations of Step 2 fail to have a conserved motif. This allows us to quickly identify and remove them.
4. The motifs generated in Step 3 were used by the MAST (Bailey and Gribskov, 1998) program to scan all the sequences in Swiss-Prot/TrEMBL (Boeckmann et al., 2003), a comprehensive non-redundant collection of known protein sequences.
5. All sequences from Step 4 that have significant scores are added to the original set of legume-specific sequences.
6. Steps 3–5 are repeated until (i) no more significant hits are found, or (ii) the sequences from Step 5 have clearly begun to dominate the PSSM (in which case further iterations would not help us with our initial set).

The procedure outlined above yielded interesting results for many of the groups that had no significant BLAST homology. For the nodule-specific CCP group, three motifs with 20, 14, and 15 amino acids respectively were identified (the first motif was the signal peptide). The first round of motif scans yielded strong hits to one sequence from *Galega orientalis*, five each from *Pisum sativum* and *Vicia faba*, and a sequence from *Trifolium replens*. The second round yielded one more strong hit to a fragmentary sequence from pea (labeled Nodulin-14), and moderate hits to a half dozen scorpion potassium-channel blocking neurotoxins.

For the seed-specific CCP group, the first round yielded a single 24-amino acid motif shared with a protease inhibitor from pear. Adding this single sequence yielded about a hundred significant hits to plant protease inhibitors, plant and insect defensins, and scorpion sodium-channel blocking neurotoxins. There are 3D structures for representatives of each of these categories of hits, and they are all known to share a common knottin fold (Thomma et al., 2002).

Other groups with no BLAST homology were identified that did have conserved motifs. One group shared domains with several Arabidopsis and Flaveria "ZF-HD homeobox proteins". A second shared a few motifs with B-like cyclins from *Medicago sativa* and *M. varia*. A third group shared three long motifs common to Globulin-1/Vicilin storage proteins in wheat, most nuts, oily seeds and legumes.

## 6. FUTURE GOALS

Using the available sequence databases and various computational algorithms, we have identified approximately 2,500 genes from *M. truncatula*, *Glycine max/soja* and *L. japonicus* that appear to be legume-specific. Currently, cDNA clones of the legume-specific genes from Medicago are undergoing full-length sequencing, which could help in further elucidating their function. In addition, sequencing of the *M. truncatula* genome will also provide information about the location and organization of these genes in the genome and can also be used to identify elements in the promoters that could determine the expression patterns of these genes.

Our next step is to confirm that the identified legume-specific genes are in fact legume-specific. This will be done using DNA from various plant species that aren't well represented in the sequence databases, including divergent legume species and taxa that are sister to the legumes (Doyle and Luckow, 2003). Finally, we hope to use molecular techniques to further characterize the expression of a subset of these genes and to determine their function in legumes.

## 7. REFERENCES

Altschul, S.F., Madden, T.L., Schaffer, A.A., Zhang, J., Zhang, Z., Miller, W., and Lipman, D.J., 1997, Gapped BLAST and PSI-BLAST: a new generation of protein database search programs, *Nucl. Acids Res.* **25**:3389–3402.

Attwood, T.K., Bradley, P., Flower, D.R., Gaulton, A., Maudling, N., Mitchell, A.L., Moulton, G., Nordle, A., Paine, K., Taylor, P., Uddin, A., and Zygouri, C., 2003, PRINTS and its automatic supplement, preprints, *Nucl. Acids Res.* **31**:400–402.

Bailey, T.L., and Elkan, C., 1994, Fitting a mixture model by expectation maximization to discover motifs in biopolymers, *Proc. Int. Conf. Intell. Syst. Mol. Biol.* **2**:28–36.

Bailey, T.L., and Gribskov, M., 1998, Combining evidence using p-values: application to sequence homology searches, *Bioinformatics* **14**:48–54.

Bateman, A., Birney, E., Cerruti, L., Durbin, R., Etwiller, L., Eddy, S.R., Griffiths-Jones, S., Howe, K.L., Marshall, M., and Sonnhammer, E.L.L., 2002, The Pfam Protein Families Database, *Nucl. Acids. Res.* **30**:276–280.

Bevan, M., Mayer, K., White, O., Eisen, J.A., Preuss, D., Bureau, T., Salzberg, S.L., and Mewes, H.W., 2001, Sequence and analysis of the *Arabidopsis* genome. *Curr Opin Plant Biol* 4(2): 105–10.

Boeckmann, B., Bairoch, A., Apweiler, R., Blatter, M., Estreicher, A., Gasteiger, E., Martin, M.J., Michoud, K., O'Donovan, C., Phan, I., Pilbout, S., and Schneider, M., 2003, The SWISS-PROT protein knowledgebase and its supplement TrEMBL in 2003, *Nucl. Acids Res.* **31**:365–370.

Doyle, J.J., and Luckow, M.A., 2003, The rest of the iceburg: legume diversity and evolution in a phylogenetic context, *Plant Physiol.* **131**:900–910.

Fedorova, M., van de Mortel, J., Matsumoto, P.A., Cho, J., Town, C.D., VandenBosch, K.A., Gantt, J.S., and Vance, C.P., 2002, Genome-wide identification of nodule-specific transcripts in the model legume *Medicago truncatula*, *Plant Physiol.* **130**:519–537.

Frühling, M., Albus, U., Hohnjec, N., Geise, G., Pühler, A., and Perlick, A.M., 2000, A small gene family of broad bean codes for late nodulins containing conserved cysteine clusters, *Plant Sci* **152**: 67–77.

Goff, S.A., Ricke, D., Lan, T.H., Presting, G., Wang, R., Dunn, M., Glazebrook, J., Sessions, A., Oeller, P., Varma, H., Hadle, D., Hutchison, D., Martin, C., Katagiri, F., Lange, B.M., Moughamer, T., Xia, Y., Budworth, P., Zhong, J., Miguel, T., Paszkowski, U., Zhang, S., Colbert, M., Sun, W.L., Chen, L., Cooper, B., Park, S., Wood, T.C., Mao, L., Quail, P., Wing, R., Dean, R., Yu, Y., Zharkikh, A., Shen, R., Sahasrabudhe, S., Thomas, A., Cannings, R., Gutin, A., Pruss, D., Reid, J., Tavtigian, S., Mitchell, J., Eldredge, G., Scholl, T., Miller, R.M., Bhatnagar, S., Adey, N., Rubano, T., Tusneem, N., Robinson, R., Feldhaus, J., Macalma, T., Oliphant, A., and Briggs, S., 2002, A draft sequence of the rice genome (*Oryza sativa* L. ssp. japonica), *Science* **296**:92–100.

Györgyey, J., Vaubert D., Jiménez-Zurdo, J., Charon, C., Troussard, L., Kondorosi, A., and Kondorosi, E., 2000, Analysis of *Medicago truncatula* nodule expressed sequence tags, *Mol. Plant-Microbe Interact.***13**:62–71.

Haft, D.H., Selengut, J.D., and White, O., 2003, The TIGRFAMs database of protein families, *Nucl. Acids. Res.* **31**:371–373.

Kaijalainen, S., Schroda, M., and Lindström, K., 2002, Cloning of nodule-specific cDNAs of *Galega orientalis*, *Physiol. Plant.* **114**: 588–593.

Lamblin, A.F., Crow, J.A., Johnson, J.E., Silverstein, K.A., Kunau, T.M., Kilian, A., Benz, D., Stromvik, M., Endre, G., VandenBosch, K.A., Cook, D.R., Young, N.D., and Retzel, E.F., 2003, MtDB: a database for personalized data mining of the model legume *Medicago truncatula* transcriptome, *Nucl. Acids. Res.* **31**:196–201.

Lee, Y., Sultana, R., Pertea, G., Cho, J., Karamycheva, S., Tsai, J., Parvizi, B., Cheung, F., Antonescu, V., White, J., Holt, I., Liang, F., and Quackenbush, J., 2002, Cross-referencing eukaryotic genomes: TIGR Orthologous Gene Alignments (TOGA), *Genome Res* **12**:493–502.

Letunic, I., Goodstadt, L., Dickens, N.J., Doerks, T., Schultz, J., Mott, R., Ciccarelli, F., Copley, R.R., Ponting, C.P., and Bork, P., 2002, Recent improvements to the SMART domain-based sequence annotation resource, *Nucl. Acids. Res.* **30**:242–244.

*Medicago truncatula* Consortium: Medicago as the nodel species for comparative and functional genomics, 2000, Minnesota, USA (January, 2003): http://www.medicago.org.

Mergaert, P., Nikovics, K., Kelemen, Z., Maunoury, N., Vaubert, D., Kondorosi, A., and Kondorosi, E., 2003, A novel family in *Medicago truncatula* consisting of more than 300 nodule-specific genes coding for small, secreted polypeptides with conserved cysteine motifs, *Plant Physiol.* **132**:in press.

Mulder, N.J., Apweiler, R., Attwood, T.K., Bairoch, A., Barrell, D., Bateman, A., Binns, D., Biswas, M., Bradley, P., Bork, P., Bucher, P., Copley, R.R., Courcelle, E., Das, U., Durbin, R., Falquet, L., Fleischmann, W., Griffiths-Jones, S., Haft, D., Harte, N., Hulo, N., Kahn, D., Kanapin, A., Krestyaninova, M., Lopez, R., Letunic, I., Lonsdale, D., Silventoinen, V., Orchard, S.E., Pagni, M., Peyruc, D., Ponting, C.P., Selengut, J.D., Servant, F., Sigrist, C.J.A., Vaughan, R., and Zdobnov, E.M., 2003, The InterPro Database, 2003 brings increased coverage and new features, *Nucl. Acids. Res.* **31**: 315–318.

National Center for Biotechnology Information (NCBI), Maryland, USA (January-March, 2003): http/www.ncbi.nih.nlm.gov.

Falquet, L., Pagni, M., Bucher, P., Hulo, N., Sigrist, C.J.A., Hofmann. K., and Bairoch, A., 2002, The PROSITE database, its status in 2002, *Nucl. Acids Res.* **30**:235–238.

Quackenbush, J., Cho, J., Lee, D., Liang, F., Holt, I., Karamycheva, S., Parvizi, B., Pertea, G., Sultana, R., and White, J., 2001, The TIGR Gene Indices: analysis of gene transcript sequences in highly sampled eukaryotic species. *Nucl. Acids. Res.* **29**:159–164.

Scheres, B., van Engelen, F., van de Knaap, E., van de Wiel, C., van Kammen, A., and Bisseling T., 1990, Sequential induction of nodulin gene expression in the developing pea nodule. *Plant Cell* **2**: 687–700

The Arabidopsis Genome Initiative, 2000, Analysis of the genome sequence of the flowering plant *Arabidopsis thaliana*, *Nature* **408**:796–815.

The Institute for Genomic Research (TIGR), Maryland, USA (January 1, 2003); http://www.tigr.org.

Thomma, B.P.H.J., Cammue, B.P.A., and Thevissen, K., 2002, Plant defensins, *Planta* **216**:193–202.

Silverstein, K.A.T., Kilian, A., Freeman, J.L., and Retzel, E.F., 2000, PANAL: an integrated resource for Protein sequence ANALyis. *Bioinformatics* **16**: 1157–1158.

Vance, C.P., Graham, P.H., and Allan, D.L., 2000, Biological nitrogen fixation: phosphorus- a critical future need?, in: *Nitrogen Fixation from Molecules to Crop Productivity*, F.O. Pederosa, M. Hungria, M.G. Yates, W.E. Newton, eds, Kluwer Academic Publishers, Dordrecht, The Netherlands, pp 509–518.

VandenBosch, K.A., and Stacey, G.A., 2003, Summaries of legume genomics projects from around the globe. Community Resources for Crops and Models, *Plant Physiol.* **131**:840–865.

Yuan, Q., Ouyang, S., Liu, J., Suh, B., Cheung, F., Sultana, R., Lee, D., Quackenbush. J., and Buell, C.R., 2003, The TIGR rice genome annotation resource: annotating the rice genome and creating resources for plant biologists, *Nucl. Acids Res.* **31**:229–233.

# Abstracts

# BRAHMS and BeerGenes: Information Management for Genetic Research on Barley and Oat

Jean Gerster, Nicholas A. Tinker,
Yella Jovich-Zahirovich, Anissa Lybaert, Shaolin Liu,
Stephen J. Molnar, and Diane E. Mather

We have developed a relational database, BRAHMS (Bioinformatic Resource for Avena and Hordeum Metabolic Sequences), to support an inter-institutional research project on sequence-based marker development for oat (*Avena sativa* L.) and barley (*Hordeum vulgare* L.) and a web interface, BeerGenes (http://gnome.agrenv.mcgill.ca/BG/), to provide public access to a subset of the data contained in the BRAHMS database. Our research integrates sequence and metabolic pathway data obtained from public databases with sequences obtained from libraries and PCR experiments in our laboratories in order to develop genetic markers that reflect differences in genes that may influence grain quality in oat and barley. BRAHMS houses sequence information from public repositories such as GenBank, as well as sequences and mapping data derived from our project research. We regularly download the current contents of public nucleotide and protein databases and conduct local BLAST homology searches. We then incorporate the results into our database. BRAHMS links DNA sequences to information about

**Jean Gerster, Yella Jovich-Zahirovich, Anissa Lybaert, Shaolin Liu, and Diane E. Mather** Department of Plant Science, McGill University, Ste-Anne-de-Bellevue, Quebec H9X 3V9, Canada. **Nicholas A. Tinker and Stephen J. Molnar** Agriculture and Agri-Food Canada, Eastern Cereal and Oilseed Research Centre, Central Experimental Farm, Ottawa, Ontario K1A 0C6, Canada.

*Genome Exploitation: Data Mining the Genome*, edited by J. Perry Gustafson, Randy Shoemaker, and John W. Snape.
Springer Science + Business Media, New York, 2005.

products for which they code (if known) and relates gene products to information about genetic loci as well as to the biochemical pathways in which they are involved. For sequences generated in the laboratory BRAHMS contains information regarding the details of the experiments from which these sequences were obtained. Information on QTLs that have been detected for grain and malt quality traits in barley are also entered into BRAHMS. At present BRAHMS contains information on 785 metabolic public web interface, BeerGenes allows any user to search for barley grain and malt quality QTLs by genomic region (BIN or BINs) or by trait. This facilitates positional comparisons of QTLs with each other and with candidate genes. As our research continues we are developing additional tools to share more of BRAHMS' capabilities with other researchers, pathways, 107 gene products, 4,700 DNA sequences, 208 QTLs and 43 other loci. A private web interface allows BRAHMS to be used in collaborative research between laboratories at McGill University and the Eastern Cereal and Oilseed Research Centre.

# Microcolinearity in the Allotetraploid *Gossypium hirsutum*

C. Grover, H. Kim, R.A. Wing, and J.F. Wendel

*Gossypium* is an interesting genus from the perspective of genome size evolution. Despite its relatively young age (5–10 million years old)[1] and conserved complement of genes, DNA content varies more than three-fold within the genus, from 2–7 pg (2C content)[2]. In an effort to determine the dynamics of genome size evolution in *Gossypium*, we have embarked on a comparative BAC sequencing project. Here we present preliminary results from the allotetraploid cotton, *Gossypium hirsutum*. BACs containing homologous genomic regions from the A genome (2C = 3.8 pg)[2] and the D genome (2C = 2 pg)[2] were randomly sheared and the fragments were cloned and sequenced. Comparisons of assembled sequences indicate a high degree of conservation between the two genomes, with fourteen genes predicted in this region, thirteen of which have moderate to high homology with annotated genes in GenBank. In addition, we uncovered three shared transposable elements, two transposable elements that are unique to the D genome, and one that is unique to the A genome. Numerous other indels distinguish these genomes, including the differential insertion of a fragment of the chloroplast gene *ycf2*. In contrast to expectations based on genome sizes, there is no size divergence in this specific 100 kb+ region, with less than 1kb difference due to indels. Future plans include extending our analysis to the model diploid parents, phylogenetic outgroups, and other genomic regions.

**C. Grover and J.F. Wendel**     Department of Botany, Iowa State University, Ames, Iowa.
**H. Kim and R.A. Wing**     University of Arizona Genomics Institute and Computational Biology, Tucson, Arizona.

*Genome Exploitation: Data Mining the Genome*, edited by J. Perry Gustafson, Randy Shoemaker, and John W. Snape.
Springer Science + Business Media, New York, 2005.

# REFERENCES

Cronn, R. C., Small, R. L., Haselkorn, T., and Wendel, J. F., 2002, Rapid diversification of the cotton genus (*Gossypium*: Malvaceae) revealed by analysis of sixteen nuclear and chloroplast genes. *American Journal of Botany*.

Wendel, J. F., Small, R. L., Cronn, R. C., and Brubaker, C. L., 1999, Genes, jeans, and genomes: reconstructing the history of cotton, In; Plant evolution in man-made habitats, L. W. D. van Raamsdonk and J. C. M. den Nijs eds., Proceedings of the VIIth International Organization of Plant Biosystematists. Hugo de Vries Laboratory, Amsterdam, The Netherlands. pp. 133–159.

# Determining the Chromosomal Location of the Wheat Leaf Rust Resistance Gene *LrW1*

C. Hiebert, J. Thomas, and B. McCallum

Leaf rust, caused by *Puccinia triticina* Eriks., is a disease of wheat (*Triticum aestivum* L.) that causes significant annual yield loss. Host genetic resistance has been successfully implemented as a disease control strategy. However, pathogen evolution results in virulent races rendering many resistance genes ineffective over time. Therefore new sources of resistance must be identified and characterized. One critical step in characterization is genetic mapping of new leaf rust resistance genes. We studied a previously identified gene temporarily named *LrW1*. To assign *LrW1* to a chromosome we used haploid deficiency mapping. We generated putative aneuploid hybrids by crossing a haploid isogenic line carrying *LrW1* (n) with a leaf rust susceptible pollinator (2n). Hybrids were inoculated at the two leaf stage with *P. triticina* virulence phenotype MBDS. The hybrids were resistant to leaf rust unless the resistant parent failed to transmit the chromosome containing *LrW1*. Leaf rust susceptible hybrids were identified and chromosome deficiencies determined with microsatellites. Three susceptible hybrids were found and each was deficient for most of chromosome 5B. Using an $F_2$ population segregating for *LrW1*, a microsatellite on chromosome $5B^S$ was shown to be linked to *LrW1* (14.1 cM). We concluded that *LrW1* is located on chromosome $5B^S$.

**C. Hiebert, J. Thomas, and B. McCallum**    Agriculture and Agri-Food Canada Cereal Research Centre, Winnipeg, Manitoba, Canada R3T 2M9.

*Genome Exploitation: Data Mining the Genome*, edited by J. Perry Gustafson, Randy Shoemaker, and John W. Snape.
Springer Science + Business Media, New York, 2005.

# Characterization of EST Libraries from Drought-Stressed Leaves and Penetrated Roots of Rice

Md S. Pathan, William G. Spollen, Mark Fredricksen, Hans J. Bohnert, Deshui Zhang, and Henry T. Nguyen

Rice (*Oryza sativa* L.) genotypes, IR62266-42-6-2 and CT9993-5-10-1-M, are well documented for their high osmotic adjustment and root penetration ability, respectively, traits which confer tolerance to drought. We constructed separately a subtracted and an un-subtracted cDNA library from drought-stressed leaf tissues of IR62266, an *indica* rice. A third library was constructed from penetrated root tissues of CT9993, a *japonica* rice. All the libraries were sequenced from the 5′ end. The combination of the two drought-stressed leaf libraries had 1911 unique genes out of 2472 ESTs. Of the unique sequences, 354 were novel as determined by BLAST search of GenBank. Some sequences showed similarities with cDNAs of known stress-related proteins, including catalase, glyceraldehydephosphate dehydrogenase, and a putative phytochrome-associated protein. Out of the 963 ESTs in the CT9993 library, 670 were unique, and 28 of these were novel. Several clones in the CT9993 library showed homology with cDNAs coding for ascorbate peroxidase, peroxidase, xyloglucan endotransglycosylase, S-adenosyl methionine

**Md S. Pathan, William G. Spollen, and Henry T. Nguyen**    Department of Agronomy, University of Missouri, Columbia, Missouri 65201.    **Mark Fredricksen and Hans J. Bohnert**    Department of Plant Biology and Department of Crop Sciences, University of Illinois, Urbana, Illinois 61801. **Deshui Zhang**    Department of Biology, Texas Tech University, Lubbock, Texas 79401.

*Genome Exploitation: Data Mining the Genome*, edited by J. Perry Gustafson, Randy Shoemaker, and John W. Snape.
Springer Science + Business Media, New York, 2005.

synthase, and other stress-related proteins. The association of the predicted gene product functions with possible mechanisms of OA and root penetration will be considered.

# The Phylogenetic Utility of *N*-Length DNA Strings in Plants

Ryan Rapp and J. Gordon Burleigh

The advent of the genomic era has produced a large amount of DNA sequence data from many organisms. Unfortunately, much of the sequence data are not useful for traditional phylogenetic analysis based on comparisons of homologous loci. We examine a method for utilizing this genomic information to resolve ancient evolutionary divergences using distance matrices based on the relative frequencies of DNA strings of length *n*. This method allows one to incorporate information from all non-repetitive DNA sequences into a phylogenetic analysis, and it avoids errors associated with mistaken sequence alignments and orthology. However, the utility of n-length string frequencies as a phylogenetic character is largely unknown, especially in plants. We attempt to address the following questions: 1) Is the frequency of *n*-length DNA strings phylogenetically informative in plants? 2) What is the basis of the phylogenetic signal within the data? 3) What are the appropriate statistical assays to evaluate the phylogenetic signal? We present the novel program, Fingerprint, to construct distance matrices based on *n*-length strings, and implement it to show the presence of a phylogenetic signal within land plants. We propose explanations for this signal and describe a new statistical methodology to analyze the robustness of the data.

**Ryan Rapp**    Department of Botany, Iowa State University, Ames, Iowa.    **J. Gordon Burleigh**
Department of Computer Science, Iowa State University, Ames, Iowa

*Genome Exploitation: Data Mining the Genome*, edited by J. Perry Gustafson, Randy Shoemaker, and John W. Snape.
Springer Science + Business Media, New York, 2005.

# Telocentrics as a Breeding Tool in Wheat

Erica Riedel and Julian Thomas

In a highly studied species like wheat, the rate at which useful traits are discovered exceeds the ability of a breeding program to introduce new variability. This is especially true where the new allele is difficult to select for. In the case of antibiotic resistance to the wheat midge, direct selection for $Sm1$ requires close scrutiny of wheat heads in the presence of a heavy midge infestation while indirect selection, using DNA markers, is possible only where there is suitable polymorphism. Even where feasible, either method is costly enough to restrict the size of the population that can be processed. As a cost effective alternative, we have introduced a relevant telocentric into elite populations segregating for the trait. Since $Sm1$ is located on $2B^S$, all the euploid progeny of a hybrid between telo $2B^L$ and a wheat midge resistant line will necessarily be resistant. Data on the transmission of $2B^L$ through ovules ($\sim$50%), through pollen ($\sim$25%) and on its effect on seed yield (low in the ditelo) show that, even with no selection, homozygous resistant types will dominate such segregating populations by F6 ($>$90%).

**Erica Riedel and Julian Thomas**  Cereal Research Center, Winnipeg, Manitoba, R3T 2M9, Canada.

*Genome Exploitation: Data Mining the Genome*, edited by J. Perry Gustafson, Randy Shoemaker, and John W. Snape.
Springer Science + Business Media, New York, 2005.

# Comparative DNA Sequence Analysis of Wheat and Rice Genomes

Mark E. Sorrells and Mauricio La Rota

The use of DNA sequence-based comparative genomics for evolutionary studies and for transferring information from model species to related large-genome species has revolutionized molecular genetics and breeding strategies for crop improvement. In this study, 2835 ESTs that have been physically mapped using wheat (*Triticum aestivum* L.) deletion lines and segregating populations were compared to the public rice (*Oryza sativa* L.) genome sequence data from 2251 ordered BAC/PAC clones using NCBI BLAST. A rice genome view of homologous wheat genome locations shows strong similarities between the previously published comparative maps based on RFLPs and the DNA sequence-based comparative map, but at a much higher resolution revealing numerous discontinuities. The physical locations of non-conserved regions do not seem to be conserved across all rice chromosomes. Several wheat ESTs having multiple wheat genome locations seem to be associated with the non-conserved regions of similarity between rice and wheat. The inverse view showing the relationship between the wheat deletion map and rice genomic sequence location revealed extensive conservation of gene content and order at the resolution conferred by the chromosome deletion breakpoints in the wheat genome. However, using only single copy genes, deletion bins in the most conserved regions often contained sequences from more

**Mark E. Sorrells and Mauricio La Rota**  Department of Plant Breeding, Cornell University, Ithaca, New York 14853.

*Genome Exploitation: Data Mining the Genome*, edited by J. Perry Gustafson, Randy Shoemaker, and John W. Snape.
Springer Science + Business Media, New York, 2005.

than one rice chromosome. This suggested that there has been an abundance of rearrangements, insertions, deletions, and duplications that, in many cases, will complicate the use of rice as a model for cross-species transfer of information in non-conserved regions.

# Marker Assisted Backcrossing—Some Lessons from Simulation

Julian Thomas and Daryl Somers

Marker assisted backcrossing accelerates restoration of the background of the recurrent parent by selecting an individual which has inherited the target trait from the non-recurrent parent, but which also has fixed more marker alleles from the recurrent parent than are expected on average. Obviously, a polyploid or large genome, in which total recombination length is high, will require more markers for good genome coverage than a small or diploid genome. At the same time however, spacing the markers closer than 50 cM will yield increasingly redundant information as linkage comes into play. We present the results of simulated experiments in which all markers were independent (consistent with spacing at about 50 cM intervals) with the number of markers ranging from 10 ($\sim$500 cM) to 100 ($\sim$5,000 cM). These simulations lead to three general conclusions. Firstly, marker assisted backcrossing works much better for small genomes than for large ones. Not only are fewer markers required but also, on average, the restitution of the best individual is more complete. Secondly, we found that increasing the number of individuals examined was an inefficient strategy for increasing the expected rate of restitution for large genomes. Typically, average restitution of the recurrent genotype was increased by about 1 marker per doubling of the population starting at 25 individuals and ending at 3200. Thirdly, the degree of restitution obtained in a single trial was quite variable compared to that observed on average. This

**Julian Thomas and Daryl Somers**    Cereal Research Center, Winnipeg, Manitoba, R3T 2M9, Canada.

*Genome Exploitation: Data Mining the Genome*, edited by J. Perry Gustafson, Randy Shoemaker, and John W. Snape.
Springer Science + Business Media, New York, 2005.

means that an observed restitution might easily be noticeably better or noticeably worse than expected. Predictions from the model were in reasonable agreement with the degree of restitution recorded in an actual backcross experiment in wheat.

# *Ceregenedb*: A Database of Coding Sequence Conservation Between Rice and Nonrice Cereals and *Arabidopsis*

Shibo Zhang, Brian C. Thomas, and Peggy G. Lemaux

The completed draft sequence of the rice (*Oryza sativa* L.) genome from both subspecies, *indica* and *japonica*, has provided the first platform in the grass family from which to analyze conservation of individual genes at both the DNA and protein levels. Because of the unavailability of full-genome sequence from other cereals, we used the UniGene sets of three nonrice cereals (barley, 6,965 sets in total; wheat, 12,467; maize, 9,897) and *Arabidopsis* (26,792) available in NCBI (http://www.ncbi.nlm.nih.gov/UniGene/) as their coding sequence representatives, comparing them with rice genome and protein sequences. From each species, there are two types of Unigene sets: CDS-type with full-length or nearly full-length mRNA sequences and EST-type with about 400-600 bp of expressed sequence from a gene. Each UniGene set sequence was analyzed for conservation with rice at both DNA and protein levels, using BLASTn against the rice *indica* genome sequence from BGI (http://btn.genomics.org.cn/rice/index.php) and BLASTx against the rice protein database in GRAMENE (http://www.gramene.org/perl/protein search), respectively. Based on current data, analysis of our results indicates that there are ~10% of CDS-type UniGene sets from each nonrice cereal that

**Shibo Zhang and Peggy G. Lemaux**    Department of Plant and Microbial Biology, University of California, Berkeley, California 94720.    **Brian C. Thomas**    College of Natural Resources Genomics Facility, University of California, Berkeley, California 94720.

*Genome Exploitation: Data Mining the Genome*, edited by J. Perry Gustafson, Randy Shoemaker, and John W. Snape.
Springer Science + Business Media, New York, 2005.

do not have identified rice homologues; most of these sequences are related to seed proteins, seed development, and resistance of the seed to insect
and fungi. From Arabidopsis, there are ~30% of CDS-type sequences that
do not have identified rice homologues. A specialized, searchable database,
*CereGeneDB* (http://genomics.enr.berkelev.edu.BarlevTag.unigene result.pl) has
been established that contains the results of the blast comparisons described
above. The data are stored in a SQL-based database, and a web interface
(http://genomics.cnr.berkeley.edu/BarleyTag/unigene_result.pl) was developed to
aid in searching the results from the database. Its availability will facilitate making
detailed comparisons of the protein and DNA data available for these plant species.
Queries can be performed using various options, including species, percent identity, length of a match, sequence type (CDS or EST), or by key word. The database
will be continuously updated as additional sequence information becomes available. We believe this database will be a useful resource for the research community
to study comparative and evolutionary genomics in the grass family and between
monocots and dicots and to aid in cloning genes, using rice as a reference.

# Index